More information about this series at https://link.springer.com/bookseries/558

Dun Deng · Mingming Liu ·
Dag Westerståhl · Kaibo Xie
Editors

Dynamics in Logic and Language

Third Tsinghua Interdisciplinary Workshop on Logic, Language, and Meaning, TLLM 2022 Virtual Event, April 1–4, 2022 Revised Selected Papers

Editors
Dun Deng
Department of Chinese Languages and Literature
Tsinghua University
Beijing, China

Mingming Liu
Department of Foreign Languages and Literatures
Tsinghua University
Beijing, China

Dag Westerståhl
Department of Philosophy
Stockholm University
Stockholm, Sweden

Tsinghua University
Beijing, China

Kaibo Xie
Department of Philosophy
Tsinghua University
Beijing, China

ISSN 0302-9743 ISSN 1611-3349 (electronic)
Lecture Notes in Computer Science
ISBN 978-3-031-25893-0 ISBN 978-3-031-25894-7 (eBook)
https://doi.org/10.1007/978-3-031-25894-7

This Springer imprint is published by the registered company Springer Nature Switzerland AG
The registered company address is: Gewerbestrasse 11, 6330 Cham, Switzerland

Preface

The 'dynamic turn' in logic and language is now almost fifty years old. The mid- to late 1970s and early 1980s saw the appearance both of adaptations of logics for reasoning about programs in computer science to the setting of modal logic and Kripke semantics, such as Propositional Dynamic Logic (PDL, Pratt, Fischer and Ladner, Segerberg, and others), and of proposals in natural language semantics, such as the Discourse Representation Theory (DRT) of Kamp and the File Change Semantics of Heim, to extend 'static' truth-conditional semantics for propositions to a dynamic semantics for discourse. Of course, 'dynamic' is a vague term. Kripke style semantics, with its rich repertoire of relations between states (of information, of program execution, of belief, of common ground, ...), is well suited as a framework for describing dynamic processes. Such descriptions can still take the meaning of a sentence, classically, to be the set of states in which it is true. Kamp and Heim style semantics changes the notion of meaning itself, now viewed as an instruction for how to update the current set of states when the sentence is accepted; put differently, as a context/information change potential.

Since then, logical dynamics and dynamics in linguistic semantics have each developed into vast and fairly well-defined areas of research, largely independent of each other although there have also been points of contact. In logic, besides describing the behavior of programs, systems modeling the actions of agents based on their attitudes (knowledge, belief, goals, ...) have been developed by scholars at the interface of logic, philosophy, computer science, artificial intelligence, and social sciences. The basic setting is still modal; Dynamic Epistemic Logic (DEL) brings updating actions like announcements with various degrees of persuasive power explicitly into the object language syntax, while other approaches keep a more standard syntax but assign new dynamic meanings.

In linguistics, Kamp and Heim style dynamic semantics was originally introduced to deal with anaphora, quantification, and presupposition projection. It has since been applied to an array of linguistic phenomena, such as epistemic modals, conditionals, plurals, tense and aspect, generalized quantifiers, propositional attitudes, vagueness, and discourse relations. Other approaches to extending classical truth-conditional meaning to a dynamic setting include situation semantics, dynamic predicate logic (DPL), variants of game-theoretic semantics, and, more recently, inquisitive semantics which treats statements and questions on a par.

The workshop brought together scholars from both traditions. Due to travel restrictions caused by the corona pandemic, it was held online, on March 31 – April 4, 2022. There were 23 submitted abstracts, 15 of which were presented at the workshop, which in addition had four invited talks and two tutorials. Two of the invited papers are published in this volume, together with seven of the contributed papers, after a careful review process. We express our sincere gratitude to all the colleagues who helped reviewing the submissions.

This was the third edition of the workshop series *Interdisciplinary Workshops on Logic, Language, and Meaning* held at Tsinghua University since its successful debut in April 2019. The plan is to continue the event and keep exploring fascinating aspects of the interface between logic and language.

December 2022

Dun Deng
Mingming Liu
Dag Westerståhl
Kaibo Xie

Organization

Tutorials

Maria Aloni (ILLC, University of Amsterdam): Dynamics in language
Johan van Benthem (Stanford University and Tsinghua University), with Chenwei Shi and Lingyuan Ye (Tsinghua University): Dynamic Epistemic Logic

Invited Speakers

Maria Aloni	University of Amsterdam
Johan van Benthem	Stanford University and Tsinghua University
Hans Kamp	Stuttgart University and University of Austin, Texas
Haihua Pan	Hongkong University

Program Committee

Gennaro Chierchia	Harvard University
Dun Deng (Co-chair)	Tsinghua University
Thomas Icard III	Stanford University
Xuping Li	Zhejiang University
Jowang Lin	Academia Sinica
Fenrong Liu	Tsinghua University and University of Amsterdam
Mingming Liu (Co-chair)	Tsinghua University
Larry Moss	Indiana University Bloomington
Stanley Peters	Stanford University
Floris Roelofsen	University of Amsterdam
Martin Stokhof	University of Amsterdam and Tsinghua University
Jakub Szymanik	University of Amsterdam
Frank Veltman	University of Amsterdam
Yingying Wang	Hunan University
Dag Westerståhl (Co-chair)	Stockholm University and Tsinghua University
Yicheng Wu	Zhejiang University
Tomoyuki Yamada	Hokkaido University
Linmin Zhang	New York University Shanghai

Organizing Committee

Zhenkun Hu	Tsinghua University
Kaibo Xie (Chair)	Tsinghua University
Jialiang Yan	Tsinghua University

Sponsors

Tsinghua University - University of Amsterdam Joint Research Center for Logic, Tsinghua University, China
Department of Philosophy, Tsinghua University, China
Department of Foreign Languages and Literatures, Tsinghua University, China
Department of Chinese Languages and Literature, Tsinghua University, China
Department of Philosophy, Stockholm University, Sweden

Contents

Contents

Neglect-Zero Effects in Dynamic Semantics

Maria Aloni(✉)

ILLC & Philosophy, University of Amsterdam, Amsterdam, The Netherlands
m.d.aloni@uva.nl

Abstract. The article presents a bilateral update semantics for epistemic modals which captures their discourse dynamics [54] as well as their potential to give rise to FC inferences [58]. The latter are derived as neglect-zero effects as in [3]. Neglect-zero is a tendency in human cognition to disregard structures that verify sentences by virtue of an empty witness set. The upshot of modelling the neglect-zero tendency in a dynamic setting is a notion of dynamic logical consequence which makes interesting predictions concerning possible divergences between everyday and logico-mathematical reasoning.

Keywords: Dynamic semantics · Epistemic MIGHT · Free choice disjunction · Bilateral negation · Post-supposition

1 Introduction

In Free Choice (FC) inferences, conjunctive meanings are derived from disjunctive modal sentences contrary to the prescriptions of classical logic:

(1) Deontic FC [35]
  a. You may go to the beach or to the cinema.
  b. $\rightsquigarrow$ You may go to the beach and you may go to the cinema.

(2) Epistemic FC [58]
  a. Mr. X might be in Victoria or in Brixton.
  b. $\rightsquigarrow$ Mr. X might be in Victoria and he might be in Brixton.

[3] presented a formal account of FC inferences in a bilateral state-based modal logic (BSML). The novel hypothesis at the core of the proposal was that FC and related inferences are a straightforward consequence of a tendency in human cognition to neglect models that verify sentences by virtue of some empty configurations (*zero-models*). Using tools from team semantics [51,57], [3] showed that the tendency to neglect zero-models (*neglect-zero* tendency) derives FC inferences (when interpreting disjunctions speakers associate each disjunct with a non-empty possibility) and their cancellation under negation. The latter result relied on the adopted bilateralism, where each connective comes with an assertion and a rejection condition, and negation is defined in terms of the latter notion [31,45,49].

D. Deng et al. (Eds.): TLLM 2022, LNCS 13524, pp. 1–24, 2023.
https://doi.org/10.1007/978-3-031-25894-7_1

In this article I will present a Bilateral Update Semantics (BiUS), building on [54]'s update semantics for epistemic MIGHT, where the neglect-zero tendency is explicitly formalised in a dynamic setting. The resulting system will derive ignorance and epistemic FC inferences as neglect-zero effects, as in [3], as well as capture the dynamics of epistemic modals in discourse, as in [54].

One crucial difference with respect to classical dynamic semantics [28,29,54] concerns the treatment of negation. Like BSML, BiUS adopts a bilateral notion of negation validating double negation elimination. When extended to the first-order case, BiUS has therefore the potential to provide an account of Barbara Partee's bathroom example (as explained at the end of Sect. 3.3, example (38)). The dynamic notion of logical consequence defined by BiUS further makes interesting predictions concerning the impact of neglect-zero on everyday reasoning and its deviation from classical logic.

The next section presents static BSML and its main motivation; Sect. 3 introduces BiUS, its core results and applications; Sect. 4 discusses some potential applications of the dynamic implementation of neglect-zero to the psychology of everyday reasoning; and Sect. 5 concludes.

2 Bilateral State-Based Modal Logic (BSML)

In team semantics formulas are interpreted with respect to a set of points of evaluation (a team) rather than single points [51,57]. In a team based modal logic [39,52], teams are sets of possible worlds. BSML is a bilateral version of a team-based modal logic [3,8] where teams are interpreted as information states. Bilateralism in this context means that we model assertion and rejection conditions rather than truth. In BSML, inferences relate speech acts rather than propositions and therefore might diverge from classical semantic entailments.

- Classical modal logic: $M, w \models \phi$, where $w \in W$
- Team-based modal logic: $M, t \models \phi$, where $t \subseteq W$
- Bilateral state-based modal logic (BSML):

$$M, s \models \phi, \text{ "}\phi \text{ is assertable in information state } s\text{"}, \quad \text{with } s \subseteq W$$
$$M, s \mathrel{=\!\!|} \phi, \text{ "}\phi \text{ is rejectable in information state } s\text{"}, \quad \text{with } s \subseteq W$$

2.1 Motivation

BSML was developed as part of a larger project with the goal to arrive at a formal account of a class of natural language inferences which diverge from classical entailments but also from canonical conversational implicatures. These include ignorance inference in modified numerals [4,16,24,47] and epistemic indefinites [5,7,32] and phenomena of free choice in indefinites and disjunction [18,35,58].

Let us focus on the case of FC inferences triggered by disjunction as in (3).

(3) You may (A or B) $\rightsquigarrow$ You may A

The logical counterpart of (3) is not valid in standard deontic logic [55]:

(4) $\Diamond(\alpha \vee \beta) \rightarrow \Diamond\alpha$ [Free Choice (FC) principle]

Plainly making the FC principle valid, for example by adding it as an axiom, would not do because we would be able to derive $\Diamond b$ from any other $\Diamond a$ as shown in (5) [35].

(5)
1. $\Diamond a$ [assumption]
2. $\Diamond(a \vee b)$ [from 1, by classical reasoning]
3. $\Diamond b$ [from 2, by FC principle]

The step leading to 2 in (5) uses the following classically valid principle:

(6) $\Diamond\alpha \rightarrow \Diamond(\alpha \vee \beta)$

The natural language counterpart of (6), however, seems invalid [44]:

(7) You may post this letter $\not\rightsquigarrow$ You may post this letter or burn it.

Thus our intuitions in natural language are in direct opposition to the principles of classical logic.

Many solutions have been proposed to solve the paradox of free choice (see [41] for a recent overview). Pragmatic **neo-Gricean** solutions derive FC inferences as conversational implicatures, i.e., pragmatic inferences derived as the product of rational interactions between cooperative language users [21,27,38,46]. On this view the step leading to 3 is unjustified. In **grammatical** solutions, instead, FC inferences result from the (optional) application of covert grammatical operators [9,10,15,20]. Again the step leading to 3 is unjustified and the paradox is solved without the need to modify the logical system. **Semantic** solutions by contrast typically change the logic. FC inferences are treated as semantic entailments [2,11,25,48]. The step leading to 3 is justified, but then it is the step leading to 2 which is no longer valid [2] or transitivity fails [25].

The main hypothesis behind the BSML solution is that FC inferences are neither the result of conversational reasoning (as proposed in neo-gricean approaches) nor the effect of optional applications of grammatical operators (as in the grammatical view). Rather they are a straightforward consequence of something else speakers do in conversation. Namely, when interpreting a sentence they create structures representing reality, pictures of the world [33] and in doing so they systematically neglect structures which (vacuously) verify the sentence by virtue of some empty configuration. This tendency, which [3] calls *neglect-zero*, follows from the expected difficulty of the cognitive operation of evaluating truths with respect to empty witness sets [12].

Models which verify a sentence by virtue of some empty set are called *zero-models.* As an illustration [3] discusses the following examples:

(8) Every square is black.
  a. Verifier: [■, ■, ■]
  b. Falsifier: [■, □, ■]
  c. Zero-models: [ ]; [△, △, △]; [◇, ▲, ♠]

(9) Less than three squares are black.
  a. Verifier: [■, □, ■]
  b. Falsifier: [■, ■, ■]
  c. Zero-models: [ ]; [△, △, △]; [◇, ▲, ♠]

The interpretation of (8) and (9) leads to the creation of structures representing reality, some verifying the sentence (the models depicted in (a)), some falsifying it (the models in (b)). The neglect-zero hypothesis states that zero-models, the ones represented in (c), are usually kept out of consideration. Zero-models are neglected because they are cognitively taxing, as confirmed by findings from number cognition [12,42]. This difficulty can be argued to further explain the special status of the zero among the natural numbers [42]; the existential import effects operative in the logic of Aristotle (the inference from *every A is B* to *some A is B*); and why downward-monotonic quantifiers (e.g., *less than n squares*) are more difficult to process than upward-monotonic ones (e.g., *more than n squares*) [12]. Since empty witnesses in zero-models encode the absence of objects, they are more detached from sensory experience and therefore harder to conceive. The inference from the perception of absence to the truth of a sentence brings in additional costs, which results in a systematic dispreference for zero-models, a neglect-zero tendency. The idea at the core of [3] is that FC and related inferences, just like the Aristotelian existential import effects, are a consequence of such neglect-zero tendency assumed to be operative among language users in ordinary conversations.

Like neo-Gricean solutions, the neglect-zero approach views FC inferences as pragmatic inferences, albeit not of the conversational implicature kind. Like semantic solutions, [3] modifies classical logic, but not to derive FC inferences as semantic entailments, but rather to formally describe the pragmatic factors responsible for these inferences and isolate their impact in a rigorous way. As we will see, BSML formally defines a pragmatic enrichment function $[\ ]^+$ and generates FC inferences only for enriched $[\Diamond(a \vee b)]^+$. When enriched, this formula is no longer derivable from $\Diamond a$. Like in grammatical approaches, the paradox in (5) is then solved as a case of equivocation:

(10)
1. $\Diamond a$
2. $\Diamond(a \vee b) \not\models [\Diamond(a \vee b)]^+$
3. $\Diamond b$

[3] gives two pieces of evidence in favor of the neglect-zero solution to the paradox of FC.

Argument from Empirical Coverage. Although pragmatic, grammatical and semantic accounts can derive the basic FC inference (labeled as Narrow Scope FC below), FC sentences give rise to complex inference patterns when embedded under logical operators: FC inferences systematically disappear in negative contexts (Dual Prohibition), but are embeddable under universal quantifiers (Universal FC); furthermore they arise under double negation (Double Negation FC) and also when disjunction takes wide scope with respect to the modal operator (Wide scope FC):

(11) Narrow Scope FC [35]

a. You may go to the beach or to the cinema.
$\rightsquigarrow$ You may go to the beach and you may go to the cinema.

b. $\Diamond(\alpha \vee \beta) \rightsquigarrow \Diamond\alpha \wedge \Diamond\beta$

(12) Dual Prohibition [6]

a. You are not allowed to eat the cake or the ice-cream.
$\rightsquigarrow$ You are not allowed to eat either one.

b. $\neg\Diamond(\alpha \vee \beta) \rightsquigarrow \neg\Diamond\alpha \wedge \neg\Diamond\beta$

(13) Universal FC [13]

a. All of the boys may go to the beach or to the cinema.
$\rightsquigarrow$ All of the boys may go to the beach and all of the boys may go to the cinema.

b. $\forall x\Diamond(\alpha \vee \beta) \rightsquigarrow \forall x(\Diamond\alpha \wedge \Diamond\beta)$

(14) (Embedded) Double Negation FC [26]

a. Exactly one girl cannot take Spanish or Calculus.
$\rightsquigarrow$ One girl can take neither of the two and each of the others can choose between them.

b. $\exists x(\neg\Diamond(\alpha(x) \vee \beta(x)) \wedge \forall y(y \neq x \rightarrow \neg\neg\Diamond(\alpha(y) \vee \beta(y)))) \rightsquigarrow$
$\exists x(\neg\Diamond\alpha(x) \wedge \neg\Diamond\beta(x) \wedge \forall y(y \neq x \rightarrow (\Diamond\alpha(y) \wedge \Diamond\beta(y))))$

(15) Wide Scope FC [58]

a. Detectives may go by bus or they may go by boat.
$\rightsquigarrow$ Detectives may go by bus and may go by boat.

b. Mr. X might be in Victoria or he might be in Brixton.
$\rightsquigarrow$ Mr. X might be in Victoria and might be in Brixton.

c. $\Diamond\alpha \vee \Diamond\beta \rightsquigarrow \Diamond\alpha \wedge \Diamond\beta$

As shown in [3] these patterns are captured by the neglect-zero approach implemented in BSML. Most other approaches instead need additional assumptions as summarised in Table 1.[1,2]

Table 1. Comparison with competing accounts of FC inference

	NS FC	Dual prohib	Universal FC	Double neg FC	WS FC
Neo-Gricean	yes	yes	no	?	no
Grammatical	yes	y/n	yes	y/n	y/n
Semantic	yes	no	yes	no	no
Neglect-zero	yes	yes	yes	yes	yes

Argument from Cognitive Plausibility. Disjunction in natural language can give rise to different pragmatic effects:

(16) You may have coffee or tea.
  a. Ignorance: $\neg K\Diamond\alpha \wedge \neg K\Diamond\beta$ (speaker doesn't know which)
  b. FC inference: $\Diamond\alpha \wedge \Diamond\beta$ (you may choose which)
  c. Scalar implicature: $\neg\Diamond(\alpha \wedge \beta)$ (you may not have both)

On neo-Gricean and grammatical accounts, FC inferences and scalar implicatures are viewed as originating from a common source: Gricean reasoning or the application of covert grammatical operators, see Table 2. The experimental literature however has shown remarkable differences between FC and scalar inferences. The former are more robust and easier to process than the latter [14,53] and are acquired earlier [50].

---

[1] Among the exceptions to this claim is [25]. See [3] for a comparison.

[2] Consider approaches in the grammatical tradition. Dual Prohibition cases are not derived directly but are explained by appealing to variations of the Strongest Meaning Hypothesis [17]. To account for wide scope FC inferences, which again cannot be generated by (recursive) applications of grammatical exhaustification, different strategies must be employed (see [9,10]). As for the case of double negation FC, as discussed in detail in [26], pages 147–149, by recursive exhaustification only we cannot capture the so-called ALL-OTHERS-FREE-CHOICE inference displayed in (14). Inclusion-based grammatical accounts [9,10], given some additional assumptions about alternatives, can derive the inference for 'exactly one' sentences but need further modifications to account for similar readings in the case of sentences using 'exactly two' or higher. In a logic-based account like BSML, the ALL-OTHERS-FREE-CHOICE reading in all these variants can be captured simply by validating dual prohibition ($\neg\Diamond(\alpha\vee\beta) \rightsquigarrow \neg\Diamond\alpha\wedge\neg\Diamond\beta$) and double negation FC ($\neg\neg\Diamond(\alpha\vee\beta) \rightsquigarrow \Diamond\alpha\wedge\Diamond\beta$). The former allows us to derive the blue part in the inference below and the latter the red part:

(i) $\exists x(\neg\Diamond(\alpha(x) \vee \beta(x)) \wedge \forall y(y \neq x \rightarrow \neg\neg\Diamond(\alpha(y) \vee \beta(y)))) \rightsquigarrow$
$\exists x(\neg\Diamond\alpha(x) \wedge \neg\Diamond\beta(x) \wedge \forall y(y \neq x \rightarrow (\Diamond\alpha(y) \wedge \Diamond\beta(y))))$

.

	Processing cost	Acquisition
FC inference	Low	Early
Scalar implicature	High	Late

The neglect-zero hypothesis has the potential to arrive at a principled explanation of these differences. On this view, FC inferences are not akin to scalar implicatures. FC follows from the assumption that when interpreting sentences language users neglect zero-models. Zero-models are neglected because cognitively taxing. Thus FC inferences result from a tendency to avoid a cognitive difficulty. Their low processing cost and early acquisition are therefore expected on this view. However, the question of how to model scalar implicatures in BSML remains. In particular, if the modelling leads to correct predictions in terms of processing and acquisition. This is one of the issues left open in [3].

Table 2. Comparison with neo–Gricean and grammatical view

	Ignorance inference	FC inference	Scalar implicature
Neo-Gricean	reasoning	reasoning	reasoning
Grammatical	debated	grammatical	grammatical
Neglect-zero	neglect-zero	neglect-zero	–

2.2 BSML: Formal Definitions

The target language is the language of propositional modal logic enriched with the non-emptiness atom, NE, from team logic [57], which [3] uses to define the pragmatic enrichment function $[\,]^+$.

Definition 1 (Language). *The language L_{BSML} is defined recursively as*

$$\phi := p \mid \neg\phi \mid \phi \wedge \phi \mid \phi \vee \phi \mid \Diamond\phi \mid \mathrm{NE}$$

where $p \in PROP$, a countable set of propositional variables.

A Kripke model for L_{BSML} is a triple, $M = \langle W, R, V\rangle$, where W is a set of worlds, R is an accessibility relation on W and V is a world-dependent valuation function for the elements of $PROP$.

Formulas in the language are interpreted in models M with respect to a state $s \subseteq W$. Both support, $\models$, and anti-support, $=\!|$, conditions are specified. On the intended interpretation $M, s \models \phi$ stands for 'formula ϕ is assertable in s' and $M, s =\!| \phi$ stands for 'formula ϕ is rejectable in s', where s stands for the information state of the relevant speaker. $R[w]$ refers to the set $\{v \in W \mid wRv\}$.

Definition 2 (Semantic clauses).

$$
\begin{aligned}
M, s \models p &\text{ iff } \forall w \in s : V(w,p) = 1\\
M, s \mathrel{=\!\!|} p &\text{ iff } \forall w \in s : V(w,p) = 0\\
M, s \models \neg\phi &\text{ iff } M, s \mathrel{=\!\!|} \phi\\
M, s \mathrel{=\!\!|} \neg\phi &\text{ iff } M, s \models \phi\\
M, s \models \phi \vee \psi &\text{ iff } \exists t, t' : t \cup t' = s \;\&\; M, t \models \phi \;\&\; M, t' \models \psi\\
M, s \mathrel{=\!\!|} \phi \vee \psi &\text{ iff } M, s \mathrel{=\!\!|} \phi \;\&\; M, s \mathrel{=\!\!|} \psi\\
M, s \models \phi \wedge \psi &\text{ iff } M, s \models \phi \;\&\; M, s \models \psi\\
M, s \mathrel{=\!\!|} \phi \wedge \psi &\text{ iff } \exists t, t' : t \cup t' = s \;\&\; M, t \mathrel{=\!\!|} \phi \;\&\; M, t' \mathrel{=\!\!|} \psi\\
M, s \models \Diamond\phi &\text{ iff } \forall w \in s : \exists t \subseteq R[w] : t \neq \emptyset \;\&\; M, t \models \phi\\
M, s \mathrel{=\!\!|} \Diamond\phi &\text{ iff } \forall w \in s : M, R[w] \mathrel{=\!\!|} \phi\\
M, s \models \text{NE} &\text{ iff } s \neq \emptyset\\
M, s \mathrel{=\!\!|} \text{NE} &\text{ iff } s = \emptyset
\end{aligned}
$$

[3] adopts the standard abbreviation: $\Box\phi := \neg\Diamond\neg\phi$, and therefore derives the following interpretation for the necessity modal:

$$
\begin{aligned}
M, s \models \Box\phi &\text{ iff for all } w \in s : R[w] \models \phi\\
M, s \mathrel{=\!\!|} \Box\phi &\text{ iff for all } w \in s : \text{there is a } t \subseteq R[w] : t \neq \emptyset \;\&\; t \mathrel{=\!\!|} \phi
\end{aligned}
$$

Logical consequence is defined as preservation of support.[3]

Definition 3 (Logical consequence). $\phi \models \psi$ *iff for all* $M, s : M, s \models \phi \Rightarrow M, s \models \psi$

In this framework we can further define team-sensitive restrictions on the accessibility relation.

Definition 4 (Team-sensitive constraints on R)

- *R is indisputable in (M, s) iff for all $w, v \in s : R[w] = R[v]$*
- *R is state-based in (M, s) iff for all $w \in s : R[w] = s$*

An accessibility relation R is state-based in a model-state pair (M, s) if all and only worlds in s are R-accessible within s. An accessibility relation R is indisputable in a model-state pair (M, s) if any two worlds in s access exactly the same set of worlds according to R. Clearly if R is state-based, R is also indisputable.

[3] proposes to use these constraints to capture the difference between epistemic and deontic modal verbs. In BSML, if we adopt a state-based accessibility relation, we can capture the infelicity of so-called *epistemic contradictions* [56], while preserving the non-factivity of $\Diamond$:

[3] For a proof-theory of BSML and related systems see [8].

1. Epistemic contradiction: $\Diamond\alpha \wedge \neg\alpha \models \bot$ (if R is state-based)
2. Non-factivity: $\Diamond\alpha \not\models \alpha$

This motivates the assumption of a state-based R for epistemic modal verbs - the assertion of the epistemic possibility of a proposition conjoined with its negation as in (17) is indeed infelicitous [30,40,54,56], but not for deontic ones, which don't give rise to similar infelicities, see (18):

(17) #It might be raining but it is not raining.

(18) You may smoke, but you don't smoke.

The accessibility relation in the case of deontic modals can at most be indisputable. Assuming s represents the information state of the relevant speaker, a state-based R leads to an interpretation of the modal as a quantifier over the epistemic possibilities the speaker entertains. An indisputable R instead only means that the speaker is fully informed about R, so, if R represents a deontic accessibility relation, indisputability means that the speaker is fully informed about (or has full authority on) what propositions are obligatory or allowed, as for example is arguably the case in performative uses.

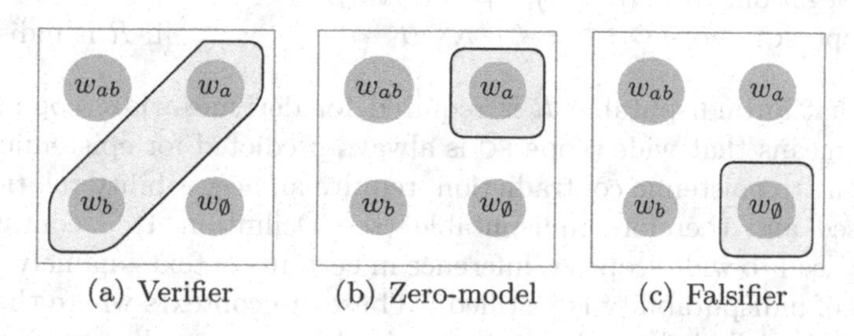

Fig. 1. Models for $(a \vee b)$.

2.3 Neglect-Zero Effects in BSML

In BSML, a state s supports a disjunction iff s is the union of two substates, each supporting one of the disjuncts. As an illustration consider the states represented in Fig. 1. In these pictures w_a stands for a world where only a is true, w_b only b, etc. The disjunction $(a \vee b)$ is supported by the first two states, but not by Fig. 3(c) because the latter consists of $w_\emptyset$, a world where both a and b are false. The state in Fig. 1(b) supports $(a \vee b)$, because we can find suitable substates supporting each disjunct: the state itself, supporting a, and the empty state, vacuously supporting b. State Fig. 1(b) is then an example of a *zero-model* for $(a \vee b)$, a model which verifies the formula by virtue of an empty witness. Using NE, [3] defines a notion of *pragmatic enrichment*, whose core effect is to disallow such zero-models. A state s supports a pragmatically enriched disjunction $[\alpha \vee \beta]^+$ iff s is the union of two *non-empty* substates, each supporting one of the disjuncts. Such enriched disjunctions thus require both their disjuncts to be live possibilities [23,58]. The pragmatic enrichment function is recursively defined for formulas in the NE-free fragment of the language as follows:

Definition 5 (Pragmatic enrichment function)

$$\begin{aligned}
[p]^+ &= p \wedge \text{NE} \\
[\neg\alpha]^+ &= \neg[\alpha]^+ \wedge \text{NE} \\
[\alpha \vee \beta]^+ &= ([\alpha]^+ \vee [\beta]^+) \wedge \text{NE} \\
[\alpha \wedge \beta]^+ &= ([\alpha]^+ \wedge [\beta]^+) \wedge \text{NE} \\
[\Diamond\alpha]^+ &= \Diamond[\alpha]^+ \wedge \text{NE}
\end{aligned}$$

The main result of this research is that in BSML $[\,]^+$-enrichments have non-trivial effects only when applied to positive disjunctions. This, in combination with the adopted notion of modality, derives FC inferences for pragmatically enriched formulas while no undesirable side effects obtain with other configurations, notably under single negation:

- Narrow scope FC: $[\Diamond(\alpha \vee \beta)]^+ \models \Diamond\alpha \wedge \Diamond\beta$
- Dual Prohibition: $[\neg\Diamond(\alpha \vee \beta)]^+ \models \neg\Diamond\alpha \wedge \neg\Diamond\beta$
- Double Negation: $[\neg\neg\Diamond(\alpha \vee \beta)]^+ \models \Diamond\alpha \wedge \Diamond\beta$
- Wide scope FC: $[\Diamond\alpha \vee \Diamond\beta]^+ \models \Diamond\alpha \wedge \Diamond\beta$ [if R is indisputable]

Notice that an indisputable R is required for deriving wide scope FC inferences. This means that wide scope FC is always predicted for epistemic modals, which, leading to epistemic contradiction, require an accessibility relation which is state-based and therefore indisputable (see Definition 4). Deontic modals instead only lead to wide scope FC inference in certain contexts, namely when the assumption of indisputability is justified.[4] These are contexts where the speaker is assumed to be fully informed about what is obligatory or allowed, for example in some performative uses of the verb. A further prediction of BSML is that cases of overt FC cancellations like (19)–(20) have to be treated as cases of wide scope FC where the assumption of indisputability is not warranted. In both cases, the prediction is arguably borne out [22,36].

(19) You may eat the cake or the ice-cream, I don't know which.

(20) You may either eat the cake or the ice-cream, it depends on what John has taken.

[4] By assuming a non-indisputable accessibility relation we can also account for the lack of FC inference in the following arguably wide scope disjunction cases discussed in [41]:

(i) a. It is OK for John to have ice-cream or it is OK for him to have cake.
b. It's conceivable that she will call or it's conceivable that she will write.

Finally notice that in this framework neglect-zero effects can be isolated and literal meanings, ruled by classical logic, can be recovered. We can indeed model the global suspension of neglect-zero effects using $\text{BSML}^{\emptyset}$, the NE-free fragment of BSML, which behaves like classical modal logic (CML).

$$\alpha \models_{BSML} \beta \text{ iff } \alpha \models_{CML} \beta \qquad [\text{if } \alpha, \beta \text{ are NE-free}]$$

In $\text{BSML}^{\emptyset}$, which captures logico-mathematical reasoning, zero-models are always allowed and play an essential role. Paraphrasing Whitehead, we can conjecture that the use of zero-models 'is only forced on us by the needs of cultivated modes of thought'.[5]

> 'The point about zero is that we do not need to use it in the operations of daily life. No one goes out to buy zero fish. It is in a way the most civilized of all the cardinals, and its use is only forced on us by the needs of cultivated modes of thought.' (A.N. Whitehead quoted by [42]).

We will return to the role of zero-models in everyday and logico-mathematical reasoning in Sect. 4.

3 An Update Semantics for Epistemic FC

In this section we introduce, BiUS, a bilateral update semantics for epistemic FC. Our point of departure is Veltman's update semantics for epistemic MIGHT [19,54]. One of the goals of [54] was to account for dynamic effects of epistemic modals in discourse. Veltman observed a difference between sequences like (21-a) and (21-b): 'it is quite normal for one's expectations to be overruled by the facts - that is what is going on in the first sequence. But once you know something, it is a bit silly to pretend that you still expect something else, which is what is going on in the second' (see [54], page 223).

(21) Veltman's sequences
  a. Maybe this is Frank Veltman's example. It isn't his example!
  b. ?This is not Frank Veltman's example! Maybe it's his example.

As we saw in the previous section, BSML can capture the infelicity of epistemic contradictions (example (17)) by modelling epistemic modals via a state-based accessibility relation. But the distinction illustrated in (21) could not be explained. BiUS will remedy to this deficiency: it will capture the discourse dynamics of epistemic modals but also their FC potential, which was not addressed in [54]:

[5] This conjecture needs to be qualified. We do engage with zero-models in our daily life, for example when interpreting sentences with downward entailing quantifiers which can only be verified by zero-models, e.g., *I have zero ideas of how to prove this* or *I went to the store to buy fish, but they didn't have any, so we'll have no fish for dinner tonight.* Downward entailing quantifiers (*no/zero*) however are more costly to process than their upward entailing counterparts (*some*), a fact which can be taken to confirm the cognitive difficulty of engaging with zero-models.

(22) Narrow scope epistemic FC

a. Mr. X might be in Victoria *or* in Brixton. $\rightsquigarrow$ Mr. X might be in Victoria *and* he might be in Brixton.

b. $\Diamond(\alpha \vee \beta) \rightsquigarrow \Diamond\alpha \wedge \Diamond\beta$

(23) Wide scope epistemic FC

a. Mr. X might be in Victoria *or* he might be in Brixton. $\rightsquigarrow$ Mr. X might be in Victoria *and* might be in Brixton.

b. $\Diamond\alpha \vee \Diamond\beta \rightsquigarrow \Diamond\alpha \wedge \Diamond\beta$

The FC inferences in (22)–(23) will be derived as neglect-zero effects as in BSML. The upshot of modelling neglect-zero in a dynamic setting is a notion of dynamic consequence which will make interesting predictions concerning possible divergences between everyday and logico-mathematical reasoning, as will be discussed in Sect. 4.

3.1 Veltman's Update Semantics for Epistemic Modals

Veltman presented an Update Semantics (US) for a propositional language with an additional unary operator (here $\Diamond$) expressing epistemic MIGHT [19,54]. The language is defined as follows precluding Boolean operations on the MIGHT-formulas.

Definition 6 (Syntax). *The languages L_{PL} and L_{US} are defined recursively as:*

$$L_{PL}: \ \alpha := p \mid \neg\alpha \mid \alpha \vee \alpha \mid \alpha \wedge \alpha$$
$$L_{US}: \ \phi := \alpha \mid \Diamond\alpha$$

where $p \in PROP$, a countable set of propositional variables.

Models are pairs $M = \langle W, V \rangle$, where W is a set of possible worlds and V is a world-dependent valuation function. Information states, $s \subseteq W$, are sets of possible worlds. Formulas in L_{US} denote functions from states to states.

Definition 7 (Updates).

$$\begin{aligned} s[p] &= s \cap \{w \in W \mid V(p,w) = 1\} \\ s[\phi \wedge \psi] &= s[\phi] \cap s[\psi] \\ s[\phi \vee \psi] &= s[\phi] \cup s[\psi] \\ s[\neg\phi] &= s - s[\phi] \\ s[\Diamond\phi] &= s, \ \textit{if } s[\phi] \neq \emptyset; \\ &= \emptyset, \ \textit{otherwise} \end{aligned}$$

Support and logical consequence are defined in terms of the update function, as standard in dynamic semantics. A state s supports a formula ϕ, $s \models \phi$, if updating s with the formula does not lead to any change.

Definition 8 (Support)

$$s \models \phi \text{ iff } s[\phi] = s$$

A formula is a consequence of a sequence of premisses, $\phi_1, \ldots, \phi_n \models \psi$ iff any state resulting from an update with the premisses, supports the conclusion.[6]

Definition 9 (Logical consequence)

$$\phi_1, \ldots, \phi_n \models \psi \text{ iff for all } s : s[\phi_1] \ldots [\phi_n] \models \psi$$

In Fig. 2 we give some illustrations for $PROP = \{a, b\}$. As in BSML, also in US the information state depicted in Fig. 2(b) is predicted to be a zero-model for $(a \vee b)$. It supports the formula by virtue of an empty witness set. Our goal here is to extend US with a notion of neglect-zero enrichments whose core effect is again to rule out such zero-models. As we saw in the previous section [3] defined pragmatic enrichment in terms of a conjunction with NE. This strategy however will not work in the context of an update semantics. A natural interpretation for an update with NE would be $s[\text{NE}] = s$, if $s \neq \emptyset$; $\emptyset$ (or undefined) otherwise. But the conjunction $\phi \wedge \text{NE}$, given Veltman's semantics for $\wedge$, would only rule out the empty state as a possible input for the formula. To model dynamic pragmatic enrichments we will instead introduce a complex expression ϕ^{NE}, where NE is interpreted as a **post-supposition**, i.e., a constraint that needs to be satisfied *after* the update with the relevant sentence:

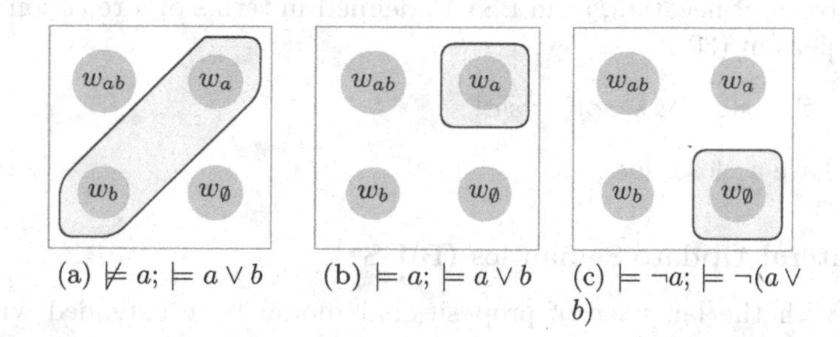

(a) $\not\models a$; $\models a \vee b$ (b) $\models a$; $\models a \vee b$ (c) $\models \neg a$; $\models \neg(a \vee b)$

Fig. 2. Illustrations of support in Veltman's update semantics

(24) $s[\phi^{\text{NE}}] = s[\phi]$, if $s[\phi] \neq \emptyset$; undefined ($\#$) otherwise

Compare the notion of a post-supposed NE with the more familiar notion of a **presupposition** $\phi_{[\psi]}$ which must be satisfied *before* the update (in the local context), and with **Veltman's** MIGHT, $\Diamond\phi$, which expresses the same non-emptiness requirement as ϕ^{NE} but differs in the produced output state: s (rather than $s[\phi]$) if the requirement is satisfied; $\emptyset$ (rather than $\#$) otherwise:

[6] It is worth mentioning that this is only one of the notions of logical consequence discussed in [54]. In fact, [54] eventually adopts a version which does not quantify over states.

- Post-supposed NE: $s[\phi^{\text{NE}}] = s[\phi]$, if $s[\phi] \neq \emptyset$; undefined (#) otherwise
- Presupposition: $s[\phi_{[\psi]}] = s[\phi]$, if $s \models \psi$; undefined (#) otherwise
- Veltman's MIGHT: $s[\Diamond\phi] = s$, if $s[\phi] \neq \emptyset$; $\emptyset$ otherwise

It is easy to see that modal disjunction (ignorance) and epistemic FC inferences are straightforwardly derived for enriched disjunctions defined in terms of post-supposed NE:

(25) $\alpha^{\text{NE}} \vee \beta^{\text{NE}} \models \Diamond\alpha \wedge \Diamond\beta$

(26) $\Diamond(\alpha^{\text{NE}} \vee \beta^{\text{NE}}) \models \Diamond\alpha \wedge \Diamond\beta$

But what about negation? Under negation (enriched) disjunction should behave classically:

(27) Mr X is not in A or B $\rightsquigarrow$ Mr X is not in A and he is not in B.

(28) Mr X cannot be in A or B $\rightsquigarrow$ Mr X cannot be in A and he cannot be in B.

Standard dynamic negation ($s[\neg\phi] = s - s[\phi]$) gives wrong results here. The formula in (29) would never be supported by any state. For example, it would be undefined in $\{w_\emptyset\}$, a state which would support $\neg(a \vee b)$:

(29) $\neg(a^{\text{NE}} \vee b^{\text{NE}})$

To fix this problem the update semantics we will introduce below adopts a bilateral notion of negation, as in BSML, defined in terms of a rejection update function $[\ ]^r$ as in (30):

(30) $s[\neg\phi] = s[\phi]^r \ \&\ s[\neg\phi]^r = s[\phi]$

Let us have a closer look.

3.2 Bilateral Update Semantics (BiUS)

We work with the language of propositional modal logic extended with ϕ^{NE}, expressing a post-supposed requirement of non-emptiness.[7]

Definition 10 (Syntax). *The language L_{BiUS} is recursively defined as:*

$$\phi := p \mid \neg\phi \mid (\phi \vee \phi) \mid (\phi \wedge \phi) \mid \Diamond\phi \mid \phi^{\text{NE}}$$

with $p \in PROP$, a countable set of propositional variables.

[7] The language of BiUS allows Boolean operations on $\Diamond$-formulas in contrast to Veltman's L_{US}, which precluded iteration and embedding of the $\Diamond$-operator. Because of this restriction, US validated idempotence ($s[\phi] = s[\phi][\phi]$) and monotonicity ($s \subseteq t$ implies $s[\phi] \subseteq t[\phi]$), which instead are not generally valid in BiUS. The adoption of a more liberal language is motivated by our linguistic goals. For example we want to explain wide scope free choice and the interpretation of *might* under negation. Some of our results however will depend on idempotence and monotonicity and, therefore, will only be valid for a fragment of the language.

Models, $M = \langle W, V \rangle$, and states, $s \subseteq W$, are defined as above. Formulas again denote functions from states to states. In Definition 11, only the clauses for post-supposed NE (clause 5) and negation (clause 6) are new. As explained above, an update with ϕ^{NE} returns the input state s updated with ϕ, if $s[\phi]$ is defined and different from $\emptyset$; undefined otherwise. Negation is defined in terms of a recursively defined rejection update function $[\ ]^r$. Notice that in the rejection update for ϕ^{NE} the contribution of NE is trivialized (clause 5′).

Definition 11 (Updates).

1. $s[p] = s \cap \{w \in W \mid V(p, w) = 1\}$
2. $s[\phi \wedge \psi] = s[\phi] \cap s[\psi]$
3. $s[\phi \vee \psi] = s[\phi] \cup s[\psi]$
4. $s[\Diamond\phi] = s$, *if* $s[\phi] \neq \emptyset$; $\emptyset$, *if* $s[\phi] = \emptyset$; *undefined (#) otherwise*
5. $s[\phi^{\text{NE}}] = s[\phi]$, *if* $s[\phi] \neq \emptyset$; *undefined (#) otherwise*
6. $s[\neg\phi] = s[\phi]^r$

where $[\phi]^r$ *is recursively defined as follows:*

1′ $s[p]^r = s \cap \{w \in W \mid V(p, w) = 0\}$
2′ $s[\phi \wedge \psi]^r = s[\phi]^r \cup s[\psi]^r$
3′ $s[\phi \vee \psi]^r = s[\phi]^r \cap s[\psi]^r$
4′ $s[\Diamond\phi]^r = s$, *if* $s[\phi]^r = s$; $\emptyset$, *if* $s[\phi]^r \neq s$; *undefined (#) otherwise*
5′ $s[\phi^{\text{NE}}]^r = s[\phi]^r$
6′ $s[\neg\phi]^r = s[\phi]$

and $s \neq x$ *means* s *is a state different from* x *(i.e., it excludes #) and* $x \cup y$ *and* $x \cap y$ *are defined only if both* x *and* y *are defined.*

Support is defined as above.

Definition 12 (Support).

$$s \models \phi \textit{ iff } s[\phi] = s$$

A formula is a consequence of a sequence of premisses, $\phi_1, \ldots, \phi_n \models \psi$ iff any state resulting from an update with the premisses, *if defined*, supports the conclusion.

Definition 13 (Logical consequence).

$$\phi_1, \ldots, \phi_n \models \psi \textit{ iff for all } s : s[\phi_1] \ldots [\phi_n] \textit{ defined } \Rightarrow s[\phi_1] \ldots [\phi_n] \models \psi$$

At last we define neglect-zero enrichments in terms of ϕ^{NE}. Pragmatically enriching an NE-free formula α, $|\alpha|^+$, consists in adding the post-supposition of NE to any subformula of α.

Definition 14 (Dynamic pragmatic enrichment). *For* NE*-free* α, $|\alpha|^+$ *defined as follows:*

$$
\begin{aligned}
|p|^+ &= p^{\text{NE}} \\
|\neg\alpha|^+ &= (\neg|\alpha|^+)^{\text{NE}} \\
|\alpha \vee \beta|^+ &= (|\alpha|^+ \vee |\beta|^+)^{\text{NE}} \\
|\alpha \wedge \beta|^+ &= (|\alpha|^+ \wedge |\beta|^+)^{\text{NE}} \\
|\Diamond\alpha|^+ &= (\Diamond|\alpha|^+)^{\text{NE}}
\end{aligned}
$$

For example, $|p \vee \neg q|^+ = (p^{\text{NE}} \vee (\neg q^{\text{NE}})^{\text{NE}})^{\text{NE}}$.

3.3 Results

It is easy to see that BiUS matches the predictions of [3] with respect to epistemic FC, while at the same time captures the discourse dynamic of epistemic MIGHT as in [54].

As in [3], we derive ignorance and FC inferences for pragmatically enriched formulas, while no undesirable side effects obtain under negation:[8,9]

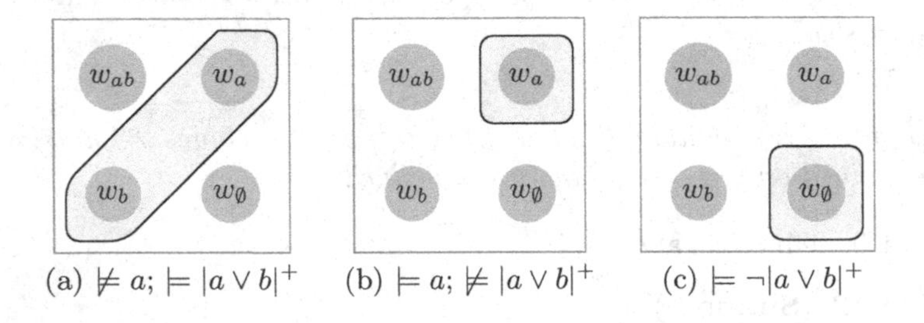

Fig. 3. Illustrations of supporting states in BiUS

1. Modal disjunction (ignorance): $|\alpha \vee \beta|^+ \models \Diamond\alpha \wedge \Diamond\beta$ (if $\alpha, \beta \in L_{US}$)

   (31) Mr. X is in Victoria or in Brixton.
   $\rightsquigarrow$ Mr. X might be in Victoria and he might be in Brixton.

2. Narrow scope epistemic FC: $|\Diamond(\alpha \vee \beta)|^+ \models \Diamond\alpha \wedge \Diamond\beta$

[8] Proofs are in appendix. See also Fig. 3 for illustrations.

[9] Notice that Modal disjunction and Negation 1 only hold for α and β of the restricted language L_{US}. Counterexamples in the unrestricted L_{BiUS} involve formulas which violate idempotence, such as epistemic contradictions. E.g., $|(p \wedge \Diamond\neg p) \vee p|^+ \not\models \Diamond(p \wedge \Diamond\neg p)$ (counterexample to Modal Disjunction), and $|\neg(\neg(p \wedge \Diamond\neg p) \vee \neg p)|^+ \not\models \neg\neg(p \wedge \Diamond\neg p)$ (counterexample to Negation 1).

(32) Mr. X might be in Victoria or in Brixton.
$\rightsquigarrow$ Mr. X might be in Victoria and he might be in Brixton.

3. Wide scope epistemic FC: $|\Diamond\alpha \vee \Diamond\beta|^+ \models \Diamond\alpha \wedge \Diamond\beta$

(33) Either Mr. X might be in Victoria or he might be in Brixton.
$\rightsquigarrow$ Mr. X might be in Victoria and he might be in Brixton.

4. Negation 1: $|\neg(\alpha \vee \beta)|^+ \models \neg\alpha \wedge \neg\beta$ (if $\alpha, \beta \in L_{US}$)

(34) Mr X is not in Victoria or in Brixton
$\rightsquigarrow$ Mr X is not in Victoria and he is not in Brixton.

5. Negation 2: $|\neg\Diamond(\alpha \vee \beta)|^+ \models \neg\Diamond\alpha \wedge \neg\Diamond\beta$

(35) Mr X cannot be in Victoria or in Brixton
$\rightsquigarrow$ Mr X cannot be in Victoria and he cannot be in Brixton.

We also match the predictions of [54] with respect to Veltman's sequences, deriving a difference between sequences $\Diamond\alpha; \neg\alpha$ and $\neg\alpha; \Diamond\alpha$ with the latter leading to contradiction and the former, instead, being consistent (update does not necessarily lead to the state of absurdity), although incoherent (no non-empty supporting state), exactly as in [54]:

(36) Veltman's sequences

a. Maybe this is Frank Veltman's example. It isn't his example!
$\Diamond\alpha, \neg\alpha \not\models \bot$

b. ?This is not Frank Veltman's example! Maybe it's his example.
$\neg\alpha, \Diamond\alpha \models \bot$

Note instead that the ordering of the conjuncts does not matter in the following versions of epistemic contradictions, which were not expressible in Veltman's original system and which are here both predicted to be incoherent:

(37) Epistemic contradictions

a. ?It might be raining but it isn't raining.
$\Diamond\alpha \wedge \neg\alpha \not\models \bot$, but incoherent

b. ?It isn't raining but it might be raining.
$\neg\alpha \wedge \Diamond\alpha \not\models \bot$, but incoherent

In fact the ϕ^{NE}-free fragment of BiUS is equivalent to US if we preclude iterations and embedding of the *might*-operator.

$$\alpha_1, \ldots, \alpha_n \models_{BiUS} \beta \text{ iff } \alpha_1, \ldots, \alpha_n \models_{US} \beta \qquad [\text{if } \alpha_1, ..., \alpha_n, \beta \in L_{US}]$$

And the non-modal and ϕ^{NE}-free fragment of BiUS is equivalent to classical logic:

$$\alpha_1, \ldots, \alpha_n \models_{BiUS} \beta \text{ iff } \alpha_1, \ldots, \alpha_n \models_{CL} \beta \qquad [\text{if } \alpha_1, \ldots, \alpha_n, \beta \in L_{PL}]$$

The latter fact means that in BiUS we can isolate neglect-zero effects just like in BSML.

In contrast to other standard dynamic systems, however, BiUS validates double negation elimination (also for non-eliminative ϕ):

- Double Negation Elimination: $\neg\neg\phi \equiv \phi$[10]

Thus BiUS also validates double negation FC (see example (14) and footnote 2 for motivation):

- Double Negation FC: $|\neg\neg\Diamond(\alpha \vee \beta)|^+ \models \Diamond\alpha \wedge \Diamond\beta$

Moreover, when applied to dynamic systems for anaphora (e.g., [29]) bilateral negation can give us a treatment of Partee's bathroom example [37]:

(38) a. Either there is no bathroom in this house or it's in a funny place.
b. $\neg\exists x Px \vee Qx$

In [29], example (38) is problematic because the last occurrence of x in (38-b) is not bound by $\exists x$, since dynamic negation neutralises the dynamic potential of existential quantifiers in its scope.

Let us assume an account of the existential quantifier, conjunction and disjunction as in [29] in combination with a bilateral notion of negation as in BiUS:

- $s[\exists x\phi] = \bigcup_{d\in D}(s[x/d][\phi])$
- $s[\phi \wedge \psi] = s[\phi][\psi]$
- $s[\phi \vee \psi] = \{i \in s \mid i \text{ survives in } (s[\phi] \cup s[\neg\phi][\psi])\}$
- $s[\neg\phi] = s[\phi]^r$

Then no matter what rejection clause one assumes for $\exists x$, the last occurrence of x in (38-b) would be bound by $\exists x$, as illustrated by the coloured parts in (39):

(39) $s[\neg\exists x Px] \cup s[\neg\neg\exists x Px][Qx] = s[\neg\exists x Px] \cup s[\neg\exists x Px]^r[Qx] = s[\neg\exists x Px] \cup s[\exists x Px][Qx]$

The full development of a quantified version of BiUS which can capture besides FC and related inferences also cross-sentential and donkey anaphora and their interactions with modality [1,29] must be left to future work.

4 Everyday vs Logico-Mathematical Reasoning

People often reason contrary to the prescriptions of classical logic. One hypothesis arising from this research is that at least in part the divergence between everyday and logico-mathematical reasoning might be due to a neglect-zero tendency. While zero-models tend to be neglected in conversation, they play a crucial role in logico-mathematical reasoning.

According to our hypothesis there are three kinds of reasonings [let $\alpha_1, ...\alpha_n, \beta$ range over NE-free formulas]:

[10] *Proof:* $s[\neg\neg\phi] = s[\neg\phi]^r = s[\phi]$.

1. *Zero-free reasonings:* classically valid reasonings which do not involve zero-models:
$$\alpha_1, ..., \alpha_n \models \beta \;\&\; |\alpha_1|^+, ..., |\alpha_n|^+ \models |\beta|^+$$
2. *Neglect-zero fallacies:* classically invalid reasonings which are valid if we neglect zero-models, e.g., ignorance and FC inferences:
$$\alpha_1, ..., \alpha_n \not\models \beta \;\&\; |\alpha_1|^+, ..., |\alpha_n|^+ \models |\beta|^+$$
3. *Zero-reasonings:* classically valid reasonings which rely on zero-models:
$$\alpha_1, ..., \alpha_n \models \beta \;\&\; |\alpha_1|^+, ..., |\alpha_n|^+ \not\models |\beta|^+$$

The hypothesis that zero-models are cognitively taxing leads to various predictions. For example, zero-reasonings should be harder for non-logically trained reasoners than zero-free reasonings. In what follows we discuss two examples illustrating these predictions.

Consider first **Disjunction Introduction**:

(40) A. THEREFORE, A OR B.

A rule-based theory which assumes that human reasoners apply the rules of Natural Deduction would predict that if asked to formulate conclusions from premise A reasoners should mention A OR B. Past experiments however showed that people who are not trained in logic do not spontaneously produce the disjunction [34]. Classical model-based theories of reasoning which link the difficulty of a reasoning solely to the amount of models involved in the reasoning process also fail to account for this fact [34,43]. In these theories, the premise leads to the construction of a model validating A. But, classically, any verifier of A is also a verifier of A OR B and so by employing a single model the conclusion A OR B should in principle be available to the everyday reasoner. Our neglect-zero hypothesis, instead, has a ready explanation of why this is not the case. A minimal verifier of A is also a verifier for A OR B but only if we allow the possibility of an empty witness for the second disjunct. Since a zero-model is involved we correctly predict that the inference is not spontaneously drawn. Disjunction introduction is indeed an example of a *zero-reasoning*, classically valid but relying on zero-models:

– $\alpha \models \alpha \vee \beta$, but $|\alpha|^+ \not\models |\alpha \vee \beta|^+$

Consider now the following two versions of **Disjunctive Syllogism** in which the ordering of the premises is reversed:

(41) A OR B; NOT A. THEREFORE, B.

(42) NOT A; A OR B. THEREFORE, B.

Both reasonings are classically valid:

- $\alpha \vee \beta, \neg\alpha \models \beta$
- $\neg\alpha, \alpha \vee \beta \models \beta$

But only (42) involves a zero-model. Any state resulting from an update with $\neg\alpha$, is a zero-model for the disjunction $\alpha \vee \beta$. This means that the sequence of updates $s[|\neg\alpha|^+][|\alpha \vee \beta|^+]$ is never defined, no matter what s is, leading to explosion:

- $|\neg\alpha|^+, |\alpha \vee \beta|^+ \models \bot$

If we reverse the ordering of the premises, with the disjunction first as in (41), the update is instead unproblematic and the output state supports the conclusion:

- $|\alpha \vee \beta|^+, |\neg\alpha|^+ \models |\beta|^+$ (and $\not\models \bot$)

(42) is then predicted to be harder than (41). We leave to future work the experimental testing of this prediction. Let me stress that this last prediction relies on the *dynamic* notion of logical consequence defined in Definition 13 and, therefore, if experimentally confirmed, would constitute independent motivation for the implementation of neglect-zero effects in a dynamic setting.

5 Conclusion

We presented an update semantics for epistemic modals capturing both their discourse dynamics and their potential to give rise to FC inferences. The latter were derived as neglect-zero effects as in [3]. In future work we intend to extend the system to the first order case and to further study and experimentally test its predictions on the impact of neglect-zero on reasoning and interpretation.

Acknowledgements. I would like to thank two anonymous reviewers for their insightful comments which led to substantial improvements. I am also grateful to Marco Degano for discussion and to Bo Flachs for his help with some of the proofs.

Appendix

BiUS, US and PL

Theorem 1. $\alpha_1, \ldots, \alpha_n \models_{BiUS} \beta$ *iff* $\alpha_1, \ldots, \alpha_n \models_{US} \beta$ [*if* $\alpha_1, ..., \alpha_n, \beta \in L_{US}$]

Proof. We only need to check the case of negation, i.e. show that $s[\gamma]^r = s - s[\gamma]$ for all s and $\gamma \in L_{PL}$ (recall that $\Diamond$ cannot appear in the scope of negation in L_{US}). We prove this by induction on the complexity of γ.

(i) $s[p]^r = s \cap \{w \in W \mid V(p,w) = 0\} = s - \{w \in W \mid V(p,w) = 1\} = s - W[p] = s - s[p]$
(ii) $s[\alpha \wedge \beta]^r = s[\alpha]^r \cup s[\beta]^r =_{IH} (s - s[\alpha) \cup (s - s[\beta]) = s - (s[\alpha] \cap s[\beta]) = s - s[\alpha \wedge \beta]$

(iii) $s[\alpha \vee \beta]^r = s[\alpha]^r \cap s[\beta]^r =_{IH} (s - s[\alpha]) \cap (s - s[\beta]) = s - (s[\alpha \cup s[\beta]) = s - s[\alpha \vee \beta]$
(iv) $s[\neg\alpha]^r = s[\alpha]$. Since $s[\alpha] \subseteq s$ by eliminativity, $s[\alpha] = s - (s - s[\alpha]) =_{IH} s - s[\alpha]^r = s - s[\neg\alpha]$.

Theorem 2. *$\alpha_1, \ldots, \alpha_n \models_{BiUS} \beta$ iff $\alpha_1, \ldots, \alpha_n \models_{PL} \beta$ [if $\alpha_1, \ldots, \alpha_n, \beta \in L_{PL}$]*

Proof. This follows from the fact that in $BiUS$ (just like in US, see [54], page 231), all $\alpha \in L_{PL}$ are such that for any s, $s[\alpha] = s \cap W[\alpha]$.

Ignorance and Free Choice

The proofs of the facts below use the following lemmas.

Lemma 1. *For $\alpha \in L_{BiUS}$ and NE-free, and any state s.*

(i) If $s[|\alpha|^+]$ is defined, then $s[|\alpha|^+] = s[\alpha]$
(ii) If $s[|\alpha|^+]^r$ is defined, then $s[|\alpha|^+]^r = s[\alpha]^r$

Proof. By an easy double induction on the complexity of α.

Lemma 2. *For $\alpha \in L_{US}$ and any state s.*

(i) Idempotence: $s[\alpha] = s[\alpha][\alpha]$ and $s[\neg\alpha] = s[\neg\alpha][\neg\alpha]$
(ii) Monotonicity: $s \subseteq t$ implies $s[\alpha] \subseteq t[\alpha]$
(iii) Downward closure of $\neg\alpha$: $s \subseteq t$ implies $t[\neg\alpha] = t \Rightarrow s[\neg\alpha] = s$.

Proof. These properties are consequences of the following two facts: (a) in L_{US} all *might*-formulas have the form $\Diamond\alpha$, where α is $\Diamond$-free; (b) all $\Diamond$-free α (i.e., $\alpha \in L_{PL}$) are such that for all s, $s[\alpha] = s \cap W[\alpha]$.

Lemma 3 (Eliminativity). *For $\phi \in L_{BiUS}$ and any state s.*

If $s[\phi]^{(r)}$ is defined, then $s[\phi]^{(r)} \subseteq s$

Fact 1 (Modal Disjunction). $|\alpha \vee \beta|^+ \models \Diamond\alpha \wedge \Diamond\beta$ *(if $\alpha, \beta \in L_{US}$)*

Proof. Suppose $s[|\alpha \vee \beta|^+]$ is defined. Then $s[|\alpha \vee \beta|^+] = s[|\alpha|^+] \cup s[|\beta|^+]$ with both $s[|\alpha|^+]$ and $s[|\beta|^+]$ defined and $\neq \emptyset$. By Lemma 1 we have $s[|\alpha|^+] = s[\alpha] \neq \emptyset$. From $s[|\alpha|^+] \subseteq s[|\alpha \vee \beta|^+]$ it follows $s[\alpha] \subseteq s[|\alpha \vee \beta|^+]$. By monotonicity of α (Lemma 2) we conclude $s[\alpha][\alpha] \subseteq s[|\alpha \vee \beta|^+][\alpha]$. Since $s[\alpha][\alpha] = s[\alpha] \neq \emptyset$ (by idempotence of α), we conclude $s[|\alpha \vee \beta|^+][\alpha] \neq \emptyset$. But then $s[|\alpha \vee \beta|^+] \models \Diamond\alpha$. Similarly for $\Diamond\beta$.

For a counterexample to Modal Disjunction with $\alpha \notin L_{US}$, let α be $(p \wedge \Diamond\neg p)$. Then $\{w_p, w_\emptyset\}[|(p \wedge \Diamond\neg p) \vee p|^+] = \{w_p\}$ is defined but does not support $\Diamond(p \wedge \Diamond\neg p)$. Thus $|(p \wedge \Diamond\neg p) \vee p|^+ \not\models \Diamond(p \wedge \Diamond\neg p)$.

Fact 2 (Narrow Scope FC). $|\Diamond(\alpha \vee \beta)|^+ \models \Diamond\alpha \wedge \Diamond\beta$

Proof. Suppose $s[|\Diamond(\alpha\vee\beta)|^+]$ is defined. Then $s[|\Diamond(\alpha\vee\beta)|^+] = s[\Diamond|(\alpha\vee\beta)|^+] = s \neq \emptyset$ and $s[|\alpha\vee\beta|^+] = s[|\alpha|^+]\cup s[|\beta|^+] \neq \emptyset$. It follows that $s[|\alpha|^+] \neq \emptyset \neq s[|\beta|^+]$. By Lemma 1 we conclude $s[\alpha] \neq \emptyset$. Hence $s[|\Diamond(\alpha \vee \beta)|^+][\Diamond\alpha] = s$ and thus $s[|\Diamond(\alpha\vee\beta)|^+] \models \Diamond\alpha$. Similarly for $\Diamond\beta$.

Fact 3 (Wide Scope FC). $|\Diamond\alpha \vee \Diamond\beta|^+ \models \Diamond\alpha \wedge \Diamond\beta$

Proof. Suppose $s[|\Diamond\alpha \vee \Diamond\beta|^+]$ is defined. Then $s[|\Diamond\alpha \vee \Diamond\beta|^+] = s[|\Diamond\alpha|^+ \vee |\Diamond\beta|^+] = s[|\Diamond\alpha|^+] \cup s[|\Diamond\beta|^+] = s \neq \emptyset$. Hence both $s[|\Diamond\alpha|^+]$ and $s[|\Diamond\beta|^+]$ are defined which means $s[\Diamond|\alpha|^+] = s[\Diamond|\beta|^+] = s \neq \emptyset$. It follows that $s[|\alpha|^+] \neq \emptyset \neq s[|\beta|^+]$. By Lemma 1 we conclude $s[\alpha] \neq \emptyset$. Hence $s[|\Diamond\alpha \vee \Diamond\beta|^+][\Diamond\alpha] = s$ and thus $s[|\Diamond\alpha \vee \Diamond\beta|^+] \models \Diamond\alpha$. Similarly for $\Diamond\beta$.

Fact 4 (Negation 1). $|\neg(\alpha \vee \beta)|^+ \models \neg\alpha \wedge \neg\beta$ *(if $\alpha, \beta \in L_{US}$)*

Proof. Suppose $s[|\neg(\alpha\vee\beta)|^+]$ is defined. Then $s[|\neg(\alpha\vee\beta)|^+] = s[|\alpha|^+ \vee |\beta|^+]^r = s[|\alpha|^+]^r \cap s[|\beta|^+]^r$. By Lemma 1 we have $s[|\alpha|^+]^r = s[\alpha]^r = s[\neg\alpha] \neq \emptyset$. From $s[|\neg(\alpha \vee \beta)|^+] \subseteq s[|\alpha|^+]^r$ we have then $s[|\neg(\alpha \vee \beta)|^+] \subseteq s[\neg\alpha]$. By idempotence, $s[\neg\alpha] = s[\neg\alpha][\neg\alpha]$, and by downword closure, $s[|\neg(\alpha\vee\beta)|^+] = s[|\neg(\alpha\vee\beta)|^+][\neg\alpha]$. Hence $s[|\neg(\alpha \vee \beta)|^+] \models \neg\alpha$. Similarly for $\neg\beta$.

Fact 5 (Negation 2). $|\neg\Diamond(\alpha \vee \beta)|^+ \models \neg\Diamond\alpha \wedge \neg\Diamond\beta$

Proof. Suppose $s[|\neg\Diamond(\alpha \vee \beta)|^+]$ is defined. Then $s[|\neg\Diamond(\alpha \vee \beta)|^+] = s[\Diamond|(\alpha \vee \beta)|^+]^r \neq \emptyset$. This means that $s[\Diamond|(\alpha \vee \beta)|^+]^r = s$ and so also $s[|(\alpha \vee \beta)|^+]^r = s[|\alpha|^+ \vee |\beta|^+]^r = s[|\alpha|^+]^r \cap s[|\beta|^+]^r = s$. By Lemma 1, $s[|\alpha|^+]^r = s[\alpha]^r$ and so $s \subseteq s[\alpha]^r$. By eliminativity, $s[\alpha]^r = s$ and so $s[\Diamond\alpha]^r = s$ and $s[\neg\Diamond\alpha] = s$. Hence $s[|\neg\Diamond(\alpha \vee \beta)|^+] \models \neg\Diamond\alpha$. Similarly for $\neg\Diamond\beta$.

References

1. Aloni, M.: Conceptual covers in dynamic semantics. In: Cavedon, L., Blackburn, P., Braisby, N., Shimojima, A. (eds.) Logic, Language and Computation, vol. III. CSLI, Stanford (2000)
2. Aloni, M.: Free choice, modals, and imperatives. Nat. Lang. Semant. **15**, 65–94 (2007). https://doi.org/10.1007/s11050-007-9010-2
3. Aloni, M.: Logic and conversation: The case of free choice. Semant. Pragmatics **15**(5), 1–39 (2022). https://doi.org/10.3765/sp.15.5
4. Aloni, M., van Ormondt, P.: Modified numerals and split disjunction: the first-order case (2021). Manuscript, ILLC, University of Amsterdam
5. Aloni, M., Port, A.: Epistemic indefinites crosslinguistically. In: The Proceedings of NELS 40 (2010)
6. Alonso-Ovalle, L.: Disjunction in alternative semantics. Ph.D. thesis, University of Massachusetts, Amherst (2006)
7. Alonso-Ovalle, L., Menéndez-Benito, P.: Epistemic Indefinites. Oxford University Press, Oxford (2015)
8. Anttila, A.: The logic of free choice. Axiomatizations of state-based modal logics. Master's thesis, ILLC, University of Amsterdam (2021)
9. Bar-Lev, M.E.: Free choice, homogeneity, and innocent inclusion. Ph.D. thesis, Hebrew University of Jerusalem (2018)

10. Bar-Lev, M.E., Fox, D.: Free choice, simplification, and innocent inclusion. Nat. Lang. Semant. **28**, 175–223 (2020). https://doi.org/10.1007/s11050-020-09162-y
11. Barker, C.: Free choice permission as resource sensitive reasoning. Semant. Pragmatics **3**(10), 1–38 (2010)
12. Bott, O., Schlotterbeck, F., Klein, U.: Empty-set effects in quantifier interpretation. J. Semant. **36**, 99–163 (2019)
13. Chemla, E.: Universal implicatures and free choice effects: experimental data. Semant. Pragmatics **2**(2), 1–33 (2009)
14. Chemla, E., Bott, L.: Processing inferences at the semantics/pragmatics frontier: disjunctions and free choice. Cognition **130**(3), 380–396 (2014)
15. Chierchia, G., Fox, D., Spector, B.: The grammatical view of scalar implicatures and the relationship between semantics and pragmatics. In: Maienborn, C., von Heusinger, K., Portner, P. (eds.) Semantics. An International Handbook of Natural Language Meaning. de Gruyter (2011)
16. Coppock, E., Brochhagen, T.: Raising and resolving issues with scalar modifiers. Semant. Pragmatics **6**(3), 1–57 (2013)
17. Dalrymple, M., Kanazawa, M., Kim, Y., Mchombo, S., Peters, S.: Reciprocal expressions and the concept of reciprocity. Linguist. Philos. **21**(2), 159–210 (1998). https://doi.org/10.1023/A:1005330227480
18. Dayal, V.: Any as inherently modal. Linguist. Philos. **21**, 433–476 (1998). https://doi.org/10.1023/A:1005494000753
19. van der Does, J., Groeneveld, W., Veltman, F.: An update on "Might". J. Log. Lang. Inf. **6**, 361–380 (1997). https://doi.org/10.1023/A:1008219821036
20. Fox, D.: Free choice and the theory of scalar implicatures. In: Sauerland, U., Stateva, P. (eds.) Presupposition and Implicature in Compositional Semantics, pp. 71–120. Palgrave MacMillan, Hampshire (2007)
21. Franke, M.: Quantity implicatures, exhaustive interpretation, and rational conversation. Semant. Pragmatics **4**(1), 1–82 (2011)
22. Fusco, M.: Sluicing on free choice. Semant. Pragmatics **12**(20), 1–20 (2019). https://doi.org/10.3765/sp.12.20
23. Geurts, B.: Entertaining alternatives: disjunctions as modals. Nat. Lang. Semant. **13**, 383–410 (2005). https://doi.org/10.1007/s11050-005-2052-4
24. Geurts, B., Nouwen, R.: At least et al.: the semantics of scalar modifiers. Language **83**(3), 533–559 (2007)
25. Goldstein, S.: Free choice and homogeneity. Semant. Pragmatics **12**(23), 1–47 (2019). https://doi.org/10.3765/sp.12.23
26. Gotzner, N., Romoli, J., Santorio, P.: Choice and prohibition in non-monotonic contexts. Nat. Lang. Semant. **28**(2), 141–174 (2020). https://doi.org/10.1007/s11050-019-09160-9
27. Grice, H.P.: Logic and conversation. In: Cole, P., Morgan, J.L. (eds.) Syntax and Semantics, Volume 3: Speech Acts, pp. 41–58. Seminar Press, New York (1975)
28. Groenendijk, J., Stokhof, M.: Dynamic predicate logic. Linguist. Philos. **14**(1), 39–100 (1991)
29. Groenendijk, J., Stokhof, M., Veltman, F.: Coreference and modality. In: The Handbook of Contemporary Semantic Theory, pp. 179–216. Blackwell (1996)
30. Hawke, P., Steinert-Threlkeld, S.: Informational dynamics of epistemic possibility modals. Synthese **195**(10), 4309–4342 (2018). https://doi.org/10.1007/s11229-016-1216-8
31. Incurvati, L., Schlöder, J.: Weak rejection. Australas. J. Philos. **95**, 741–760 (2017)
32. Jayez, J., Tovena, L.: Epistemic determiners. J. Semant. **23**, 217–250 (2006)

33. Johnson-Laird, P.N.: Mental Models. Cambridge University Press, Cambridge (1983)
34. Johnson-Laird, P., Byrne, R., Schaeken, W.: Propositional reasoning by model. Psychol. Rev. **99**, 418–439 (1992)
35. Kamp, H.: Free choice permission. Proc. Aristot. Soc. **74**, 57–74 (1973)
36. Kaufmann, M.: Free choice is a form of dependence. Nat. Lang. Semant. **24**(3), 247–290 (2016). https://doi.org/10.1007/s11050-016-9125-4
37. Krahmer, E., Muskens, R.: Negation and disjunction in discourse representation theory. J. Semant. **12**(4), 357–376 (1995)
38. Kratzer, A., Shimoyama, J.: Indeterminate pronouns: the view from Japanese. In: Lee, C., Kiefer, F., Krifka, M. (eds.) Contrastiveness in Information Structure, Alternatives and Scalar Implicatures. SNLLT, vol. 91, pp. 123–143. Springer, Cham (2017). https://doi.org/10.1007/978-3-319-10106-4_7
39. Lück, M.: Team logic: axioms, expressiveness, complexity. Ph.D. thesis, University of Hannover (2020)
40. Mandelkern, M.: Coordination in conversation. Ph.D. thesis, Massachusetts Institute of Technology (2017)
41. Meyer, M.C.: Free choice items disjunction – "an apple or a pear". In: The Wiley Blackwell Companion to Semantics. Wiley Blackwell (2020)
42. Nieder, A.: Representing something out of nothing: the dawning of zero. Trends Cogn. Sci. **20**, 830–842 (2016)
43. Quelhas, A., Rasga, C., Johnson-Laird, P.: The analytic truth and falsity of disjunctions. Cogn. Sci. **43**, e12739 (2019)
44. Ross, A.: Imperatives and logic. Theoria **7**, 53–71 (1941)
45. Rumfitt, I.: 'Yes and No'. Mind **109**, 781–823 (2000)
46. Schulz, K.: A pragmatic solution for the paradox of free choice permission. Synthese **142**, 343–377 (2005). https://doi.org/10.1007/s11229-005-1353-y
47. Schwarz, B.: Consistency preservation in quantity implicature: the case of at least. Semant. Pragmatics **9**(1), 1–47 (2016)
48. Simons, M.: Dividing things up: the semantics of or and the modal/or interaction. Nat. Lang. Semant. **13**(3), 271–316 (2005). https://doi.org/10.1007/s11050-004-2900-7
49. Smiley, T.: Rejection. Analysis **56**(1), 1–9 (1996)
50. Tieu, L., Romoli, J., Zhou, P., Crain, S.: Children's knowledge of free choice inferences and scalar implicatures. J. Semant. **33**(2), 269–298 (2016). https://academic.oup.com/jos/article-abstract/33/2/269/2413864?redirectedFrom=fulltext
51. Väänänen, J.: Dependence Logic. Cambridge University Press, Cambridge (2007)
52. Väänänen, J.: Modal dependence logic. In: Apt, K.R., van Rooij, R. (eds.) New Perspectives on Games and Interaction, pp. 237–254. Amsterdam University Press (2008)
53. Van Tiel, B., Schaeken, W.: Processing conversational implicatures: alternatives and counterfactual reasoning. Cogn. Sci. **105**, 93–107 (2017)
54. Veltman, F.: Defaults in update semantics. J. Philos. Log. **25**, 221–261 (1996). https://doi.org/10.1007/BF00248150
55. von Wright, G.: An Essay on Deontic Logic and the Theory of Action. North Holland (1968)
56. Yalcin, S.: Epistemic modals. Mind **116**(464), 983–1026 (2007)
57. Yang, F., Väänänen, J.: Propositional team logics. Ann. Pure Appl. Logic **168**, 1406–1441 (2017)
58. Zimmermann, E.: Free choice disjunction and epistemic possibility. Nat. Lang. Semant. **8**, 255–290 (2000). https://doi.org/10.1023/A:1011255819284

Another Look at Chinese Donkey Sentences: A Reply to Cheng and Huang (2020)

Haihua Pan(✉) and Hang Kuang

The Chinese University of Hong Kong, Shatin, N.T, Hong Kong SAR, China
panhaihua@cuhk.edu.hk, clydekuang@link.cuhk.edu.hk

Abstract. This paper re-examines two types of donkey conditional sentences in Mandarin Chinese, first identified by Cheng and Huang (1996), who argue that there is a strict correspondence between types of conditional sentences and interpretational strategies for the anaphoric elements, i.e., bare conditionals use the unselective binding mechanism of classic DRT, while *ruguo/dou*-conditionals need the E-type strategy. This claim has been challenged by Pan and Jiang (2015), who provide ample evidence against Cheng and Huang's paradigm. In their recent reply, Cheng and Huang (2020) attempt to further explain the counter-examples given by Pan and Jiang (2015), and maintain that their original proposal can be defended. We will examine Cheng and Huang's (2020) responses in detail, and argue that their defense does not stand up to scrutiny. We conclude that there is no correspondence between conditional types and interpretative mechanisms in Chinese donkey sentences. We also provide additional data concerning sentences with pairs of identical NPs, and point out that bare conditionals are one special example of a general strategy of expressing co-variation in Mandarin.

Keywords: Donkey sentences · Unselective binding · E-type

1 Introduction

In their pioneering study on donkey anaphora in Chinese, Cheng and Huang (1996, C&H hereafter) distinguish two types of donkey conditional sentences, namely, the so-called 'bare conditionals' and *ruguo/dou*-conditonals: bare conditionals do not have any overt marker to connect the antecedent clause and the consequent clause, as shown in (1a), while *ruguo*-conditionals are marked by the conditional morpheme *rugou* 'if' in the antecedent clause, as in (2a) and *dou*-conditionals are signaled by the quantifier-like element *dou* 'all' in the consequent clause, exemplified by (3a). According to C&H, both types of donkey conditionals have a *wh*-phrase in the antecedent clause, though they differ in what anaphoric elements are allowed in their consequent clause. Specifically, bare conditionals require another identical *wh*-phrase and disallow an anaphoric (overt or null) pronoun or a definite description, as illustrated in (1b), whereas *ruguo/dou*-conditionals exhibit a complementary distribution of anaphoric elements, only permitting a (overt or null) pronoun or a definite description and prohibiting a *wh*-phrase in the consequent, as shown in (2b) and (3b). C&H further claim that the two types of conditional sentences

D. Deng et al. (Eds.): TLLM 2022, LNCS 13524, pp. 25–51, 2023.
https://doi.org/10.1007/978-3-031-25894-7_2

are associated with two different interpretation strategies, i.e., the unselective binding mechanism of classic Discourse Representation Theory (cf. Kamp 1981, Heim 1982) for bare conditionals and the E-type strategy (cf. Cooper 1979, Evans 1980, Heim 1990) for *ruguo/dou*-conditionals.

(1) a. Shei xian lai, shei xian chi.
who first come who first eat
'If X comes first, X eats first.'
b. *Shei xian lai, ta/pro/na-ge ren xian chi.
who first come he/pro/that-CL person first eat
'If X comes first, X/that person eats first.'

(2) a. **Ruguo** ni kandao shei, qing jiao ta/pro/na-ge ren lai jian wo.
if you see who please tell him/pro/that-CL person come see me
'If you see someone, please ask him/that person to come see me.'
b. ***Ruguo** ni kandao shei, qing jiao shei lai jian wo.
if you see who please tell who come see me
'If you see someone, please ask him to come see me.'

(3) a. Ni jiao shei jin-lai, wo **dou** jian ta/pro/na-ge ren.
you ask who enter I all see him/pro/that-CL person
'Whoever you ask to come in, I'll see him/that person.'
b. *Ni jiao shei jin-lai, wo **dou** jian shei.
you ask who enter I all see who
'Whoever you ask to come in, I'll see him.'

The empirical basis and theoretical analysis of C&H (1996) have been challenged, notably by Pan and Jiang (1997, 2015, P&J hereafter), who point out that the identified complementary distribution is only apparent and object to the claim that a strict correspondence exists between the types of donkey conditionals and the interpretation strategies for the anaphoric elements. They argue that what C&H have observed and proposed are only the preferred/default distributional patterns, and given appropriate contexts, one can deviate from the preferred patterns. In particular, a pronoun can be used in the consequent of bare conditionals and a *wh*-phrase can occur in the consequent of *ruguo/dou*-conditionals as well; a *wh*-phrase can employ the E-type strategy and a pronoun can also be interpreted as a bound variable. A Bound Variable Hierarchy is proposed by P&J to account for the interpretative flexibility of pronouns and *wh*-phrases in Chinese.

Recently, C&H (2020) responded to the challenges brought up by P&J and addressed some of the issues raised. They concluded that their original proposal can still be maintained. In this paper, we first review the major empirical data that are in dispute from previous works and establish the correct distributional patterns of anaphoric elements in

Chinese donkey sentences. We then discuss C&H's original proposal and their recent further elaboration. A close look at C&H (2020) reveals that their responses to P&J's objections are still problematic, and that P&J's null hypothesis, i.e., there is no correspondence between sentence types and interpretation strategies, can be defended.[1]

2 The Empirical Picture

According to C&H (1996), the distributions of anaphoric elements in Chinese donkey sentences demonstrated by (1)-(3) can be summarized as in (4)-(6). Crucially, pronouns (and definite descriptions) have a complementary distribution with *wh*-phrases in the consequent clause of bare conditionals and *ruguo/dou*-conditionals.

(4) Bare conditionals
[$_{\text{antecedent}}$...*wh*...], [$_{\text{consequent}}$...*wh*/*pronoun...]

(5) *Ruguo*-conditionals
[$_{\text{antecedent}}$ *ruguo*...*wh*...], [$_{\text{consequent}}$...pronoun/**wh*...]

(6) *Dou*-conditionals
[$_{\text{antecedent}}$ ...*wh*...], [$_{\text{consequent}}$...*dou*...pronoun/**wh*...]

Based on the above distributional patterns, C&H suggest that both the unselective binding mechanism and the E-type strategy are necessary for the analysis of Chinese donkey sentences. As noted earlier, they argue that there exists a correspondence between the types of conditionals and the interpretation strategies: bare conditionals are interpreted by unselective binding, and *ruguo/dou*-conditionals employ the E-type strategy. To illustrate, the pair of identical *wh*-phrases in a bare conditional such as (7a) are analyzed as indefinites, which semantically function as individual variables, following Heim (1982), unselectively bound by a default null necessity operator, which provides the universal quantificational force, as shown in (7b). On the other hand, in both *dou*-conditionals and *ruguo*-conditionals, the *wh*-phrase in the antecedent clause is treated as an existential quantifier, and the pronoun in the consequent clause as an E-type pronoun, which stands for a definite description recoverable from the antecedent clause (cf. (8b) and (9b)). Furthermore, the antecedent clause containing a *wh*-phrase (qua an existential quantifier) in *dou*-conditionals is regarded as a *wh*-question, and *dou* 'all' quantifies over the set of propositions denoted by this question, as represented in (9b).

[1] Other approaches to Chinese donkey sentences, which target specifically bare conditionals, include the question-based approach (e.g., Liu 2018, Xiang 2020, Li 2021) and the relative-clause-based approach (e.g., Wen 1997, 1998, Huang 2010, Luo and Crain 2011, Chen 2019). The former takes the *wh*-phrase-containing clause in a bare conditional as a genuine *wh*-question, while the latter analyzes it as a relative clause. Due to space limit, we focus only on the debate between C&H and P&J here, and discuss the problems of other alternative approaches in another paper.

(7) a. Shei xian lai, shei xian chi.
who first come who first Eat
'If X comes first, X eats first.'
b. NEC_x[x comes first][x eats first]

(8) a. **Ruguo** ni kandao shei, qing jiao ta/pro/na-ge ren lai
if you see who please tell him/pro/that-CL person come
jian wo.
see me
'If you see someone, please ask him/that person to come see me.'
b. $\exists x[\text{person}(x) \wedge \text{you saw } x] \rightarrow$ you ask **the person you saw** to see me

(9) a. Ni jiao shei jin-lai, wo **dou** jian ta/pro/na-ge ren.
you ask who enter I all see him/pro/that-CL person
'Whoever you ask to come in, I'll see him/that person.'
b. $\forall p[\exists x[\text{person}(x) \wedge p = \text{you ask } x \text{ to come in}] \rightarrow$ I'll see **the person you ask** in the event of p]

P&J agree with C&H in the sense that two interpretation strategies are needed for Chinese conditionals. What they reject is the strict correspondence between conditional types and interpretation mechanisms and the complementary distribution of *wh*-phrases and pronouns in the consequent clauses of the two types of conditionals, since abundant exceptions to the patterns exhibited in (4)-(6) can be found. For P&J, the correlation proposed by C&H is only the default way of interpretating anaphoric relations in Chinese conditionals, because *wh*-phrases are more suitable to be used as variables, and pronouns are by default used as E-type pronouns in Chinese.[2] In addition, P&J point out that Chinese bare conditionals prefer unselective binding while *dou*/*ruguo*-conditionals tend to employ the E-type strategy. Since what C&H propose is just the default pattern, deviations from it are possible, though additional contextual support is needed. That is, under appropriate contexts, a pronoun can occur in the consequent of a bare conditional, and a *wh*-phrase can also appear in the consequent clause of a *ruguo*/*dou*-conditional. It is also possible for the *wh*-phrase in consequent of a bare conditional to have an E-type interpretation, and for the pronoun in the consequent clause of a *ruguo*/*dou*-conditional to be a bound variable. We take this to be the null hypothesis, namely that both types of conditionals can employ either of the two interpretation strategies and with appropriate contexts both *wh*-phrases and pronouns can be interpreted either as bound variables or as E-type pronouns.

[2] We regard E-type pronouns as pronouns that semantically stand for definite descriptions whose descriptive content can be recovered from linguistic or non-linguistic contexts, without taking a stand on the long-debated issue of whether definite NPs (and E-type pronouns) are referential or quantificational.

2.1 *Wh*/pronoun Alternations in Bare Conditionals

C&H claim that in bare conditionals, the anaphoric *wh*-phrase in the consequent clause cannot be replaced by a pronoun (cf. 1b). However, P&J demonstrate that this is not case, as examples like (10) show that *wh*-phrases can alternate with pronouns in the consequent clause.

(10) a. Shei yao zhe puo-chang, wo jiu rang gei **shei/ta**.
who want this broken.factory I then give to who/him
'Whoever wants this broken factory, I will give it to him.'

b. Shei bu dui, wo jiu shuo **shei/ta** bu dui.
who not right I then say who/he not right
'Whoever is not right, I will say he is not right.'

Lin (1996) also observes that the anaphoric element in a bare conditional does not have to a *wh*-phrase, as his examples in (11) exemplify. He further argues that a bare conditional with an anaphoric pronoun differs from the one with a *wh*-phrase in that the former only has a one-case interpretation in the sense of Kadmon (1987, 1990), whereas the latter can have either a one-case reading or a multi-case reading. Contrary to Lin, P&J point out that conditional sentences with an anaphoric pronoun in the consequent clause can also have a multi-case reading, similar to the interpretation of sentences that employ an anaphoric *wh*-phrase.[3]

(11) a. Shang ci shei mei jiang-wan, jintian jiu you **shei/ta**
last time who not talk.finish today then from who/him
xian kaishi.
fist begin
'Today we begin with the one who did not finish his talk last time.'

b. Ni zuotian gen shei yi zu, jintian ni jiu haishi
you yesterday with who one group today you then still
gen **shei/ta** yi zu.
with who/him one group
'Today you form a group with the one who was in your group yesterday.'
(Lin 1996, p.249)

In fact, C&H have noticed the possibility of using anaphoric pronouns in bare conditionals, though they claim that in all the bare conditionals that allow *wh*-phrase/pronoun alternations, there has to be a connective *jiu* 'then' in the consequent. For them, *jiu* is optional in bare conditionals, and its presence signals a covert conditional morpheme

[3] Some other authors also reported a *wh*-pronoun contrast. For instance, Wang (2006) claims that a *wh…wh* conditional has a different meaning from a *wh…pronoun* conditional: the former expresses universal quantification, i.e., the set of individuals denoted by the first *wh*-expression is a subset of the set denoted by the second *wh*-expression, while the latter states that the situation described by the second clause is a necessary consequence of the situation expressed by the first clause. It is not clear whether these two readings are really different. For us, the use of a pronoun in the consequent clause is compatible with a universal reading, too.

ruguo 'if'. This means that the sentences in (10) and (11) with pronouns in the consequent are actually not bare conditionals but *ruguo*-conditionals with a covert *ruguo*, which explains why a pronoun can be used in these sentences. However, a lot of examples can be found in which the anaphoric *wh*-phrase is interchangeable with a pronoun even in the absence of the connective *jiu*. For instance, P&J's examples in (12) have a pronoun in the consequent of bare conditionals without *jiu*. A noticeable feature of these examples is that they all seem to carry an imperative flavour (p.c. Lisa Cheng), which may be seen as a kind of additional contextual support for the use of pronouns in bare conditionals, a less preferred option, according to P&J.

(12) a. Shei yao zhe po-chang, rang gei **shei/ta**
who want this broken.factory give to who/him
hao le.
good SFP
'Whoever wants this broken-factory, give it to him then.'

b. Shei yao zhe po-chang, rang **shei/ta** dao bangonshi
who want this broken.factory let who/him to office
lai zhao wo.
come find me
'Whoever wants this broken factory, let him/her come to my office to see me.'

c. Shei xiang qu Beijing, **pro** bixu dao wo zheli baodao.
who want go Beijing pro must to me here register
'Whoever wants to go to Beijing, he must register with me.'

(P&J 2015, p.164)

2.2 *Wh*/pronoun Alternations in *Ruguo*-Conditionals and *Dou*-Conditionals

C&H claim that the anaphoric element in the consequent clause of *ruguo*-conditionals and *dou*-conditionals cannot be a *wh*-phrase. In contrast, P&J demonstrate that in both types of conditionals, the use of a *wh*-phrase in the consequent is possible, as shown in (13) for *ruguo*-conditionals and (14) for *dou*-conditionals. Other authors (e.g., Huang 2003, Wang 2006) have made similar observations.

(13) a. Ruguo shei yao zhe po-chang, jiu rang **ta/shei**
If who want this broken.factory then let him/who
dao bangongshi lai zhao wo.
to office come find me
'Whoever wants this broken factory, let him come to my office to see me.'

b. Xiangshan meiyou liang-pian xiangtong de hongye, ruguo
Xiangshan not.have two-CL same MOD red.leave if
shei zhaodao-le **ta/shei** jiu shi zui xingfu de ren.
who find-PERF he/who then be most happy MOD person
'There are no two identical maple leaves in Xiangshan. If anyone/someone finds them, then he will be the happiest person.'

(P&J 2015, p.165)

(14) a. Amei shuo: "Gei shei kan, **ta/shei** dou hui shuo wo
Amei say "give who look he/who all will say I
shi haoxinhaoyi."
be good.will
'Amei said: 'Whomever you give to look at (it), he will say that I meant well.'

(P&J 2015, p.165)

b. Zhang Manyu ai-shang shei, **ta/shei** dou hui xinxiruokuang.
Maggie.Cheung love who he/who all will extremely.happy
'Whoever Maggie Cheung falls in love with will be extremely happy.'

c. Shuo gei shei ting, **ta/shei** dou hui xiao-diao ya.
say to who liesten he/who all will laugh.fall teeth
'Whoever you tell this to, he will laugh his head off.'

(Wang 2006, p.47)

It is worth mentioning that according to Lin (1996) a *ruguo/yaoshi*-conditional allows a pronoun to be replaced by a *wh*-phrase in the consequent clause only if the *wh*-phrase in the antecedent precedes *ruguo/yaoshi* 'if'. Our examples in (13), as well as the examples in (15) below, which are naturally occurring examples extracted from corpus by Chen (2015), clearly show that *ruguo* can in fact precede the *wh*-phrase in the antecedent clause when another corresponding *wh*-phrase occurs in the consequent clause. The sentence (15c) is specially revealing, as it has both *ruguo* and *dou* co-occurring in the same sentence but still allows a *wh*-phrase in the consequent clause.

(15) a. You xiangzhen qiye jiuchang lingdao ceng xuanbu: "ruguo
have country company wine.factory leader once announce if
shei neng bangzhu qing lai Wu Ming, gei shei
who can help invite come Wu Ming give who
yi-wan yuan."
ten.thousand yuan
'A director of a wine factory of a country company once announced: Whoever help invite Wu Ming to come, we will give that person ten thousand yuan.'

b. Ruguo shei zhuazhu-le zhe yi huan, shei jiu keyi huode
if who catch-PERF this one link who then can get
keguan de huibao.
considerable MOD reward
'Whoever has caught this link can get a considerable reward.'

c. Ruguo shei qichuang gan shenme, biru dao shui he,
if who get.up do what for.example pour water drink
na yan, qu maojin, shei dou yao chuan-shang yifu.
take cigarette take towel who all should wear.up clothes
'Whoever gets up from bed to do something, like add some water to drink, get some cigarettes, or get the towel, he must put on his clothes.'
(Chen 2015, p.10-11)

2.3 Interim Summary

The empirical picture emerging from our survey of the relevant data seems to be that in a Chinese conditional sentence, if a *wh*-phrase occurs in the antecedent clause as an indefinite, it is generally possible to use another *wh*-phrase or a pronoun as the anaphoric element in the consequent clause, as schematized in (16) below, no matter what additional element there is in the sentence, be it the connective *jiu*, the conditional morpheme *ruguo/yaoshi* 'if', or the universal quantifier *dou* 'all'. Given the observations made in this section, the distinction between bare conditionals and *ruguo/dou*-conditionals becomes blurred. Irrespective of the types of conditionals, the pattern seems to be the same, i.e., (16). The question then is whether there is still a need to distinguish the two types of conditionals, if both behave in a similar way, and whether it is still true that there exists a strict correspondence between sentence types and interpretation strategies, as C&H claim.

(16) $[_{\text{antecedent}}$ ...*wh*...$]$ $[_{\text{consequent}}$ ...*wh*/pronoun...$]$

Although the *wh*/pronoun alternation is generally possible in Chinese conditionals, the paradigms identified by C&H, we believe, can only be regarded as the default patterns. In other words, a bare conditional prefers *wh*-phrases to occur in pairs, while a *ruguo/dou* conditional tends to use a pronoun in the consequent. Bare conditionals by default are interpreted via unselective binding and *ruguo/dou* conditionals prefer to adopt the E-type strategy. There is still a need to distinguish two types of conditionals, though the difference does not lie in whether they allow *wh*/pronoun alternations or not, but in the default option they take. Any deviation from the default patterns requires additional contexts, as suggested by P&J.[4] Earlier we saw that an imperative context allows a pronoun to occur more easily in the consequent of a bare conditional. Another reliable predictor of the possibility of *wh*/pronoun alternation is the presence of the connective *jiu* 'then' in the consequent clause (cf. 10 and 11).

3 Can C&H'S Original Proposal Be Maintained?

C&H (2020) attempt to address the issues raised by P&J for their original proposal concerning *wh*/pronoun alternations and the correspondence between sentence types and interpretation strategies, and conclude that the additional data presented by P&J do not actually pose a challenge to them. In this section, we will look at their responses in detail to see if their original proposal remains unchallenged.

3.1 Bare Conditionals and *Ruguo*-Conditionals

The impossibility of using a pronoun in the consequent clause of bare conditionals is explained in C&H (1996) as follows. In a tripartite quantified structure 'NECx[...x...][...x...]', if the variable in the consequent clause is realized as a pronoun, it will be bound directly by the necessity operator in an A'-position, which amounts to treating the pronoun as a resumptive pronoun. Given the assumption that there are no true resumptive pronouns in Chinese, a pronoun used here is ungrammatical. Besides, the pronoun cannot be bound by the first variable introduced by the *wh*-phrase in the antecedent clause, as the former is not accessible to the latter. Therefore, there is no way in which the pronoun can be bound, and thus it is excluded from the consequent clause.

Several problems render this account untenable. C&H's reasoning above yields wrong predictions for English conditionals. They would predict that a pronoun in the consequent clause of an English conditional sentence is also ruled out, since they assume that like Chinese, English lacks true resumptive pronouns as well. Consequently, the pronoun *it* in a sentence *If John owns a donkey, he beats it* would have to be regarded as a resumptive pronoun and thus ungrammatical. It is doubtful that the pronoun *it* here is a resumptive pronoun, and the grammaticality of such typical donkey sentences suggests that C&H's account resorting to prohibition against resumptive pronouns is not on the right track. In addition, the claim that the *wh*-phrase in the antecedent clause is not accessible to the anaphoric pronoun in the consequent clause rests upon the assumption that the pronoun is bound by the antecedent *wh*-phrase, which is not an accessible binder in the sense of Higginbotham (1980). Under the Kamp-Heim theory, however, it is the implicit necessity operator (or other adverbs of quantification, if any) that binds both variables. The classic DRT assumes that the antecedent NP is accessible to the anaphoric NP if the latter occurs within the scope of the quantifier that binds the former, so that they can be translated using the same variable (cf. also Kadmon 2001). Therefore, it is not the case that the antecedent *wh*-phrase is not accessible to the anaphoric pronoun in

[4] Huang A. (2003) argues that there is only one type of Chinese conditionals based on the assumption that a *wh*-phrase and a pronoun are freely interchangeable in the consequent clause of all conditionals, which is not the case.

a Chinese bare conditional, at least not in the sense of 'accessibility' in the Kamp-Heim theory.

Recall that in both bare and *ruguo*-conditionals, the anaphoric element in the consequent clause can take the form of either a *wh*-phrase or a pronoun, contrary to C&H's claim that bare conditionals only allow *wh*-phrases, and that *ruguo*-conditionals only permit pronouns. C&H (2020) argue that the apparent exceptions are not problematic for their proposal. As briefly noted in Sect. 2, they point out that whenever the *wh*-phrase in the consequent clause of a bare conditional can be replaced by a pronoun, the connective *jiu* 'then' must be present. They think that the presence of *jiu* is indicative of a covert *ruguo* 'if', and thus a bare conditional with *jiu* has an ambiguity: it can either be a true bare conditional or a *ruguo*-conditional with a covert *ruguo*. Consequently, a pronoun occurring in the consequent is not surprising, because with the pronoun, the sentence is essentially a *ruguo*-conditional in disguise. According to this analysis, the sentences in (10), repeated here as (17), with *wh*/pronoun alternations, can have either one of the representations in (18). When a *wh*-phrase is used as the anaphoric element, both *wh*-phrases are unselectively bound by the necessity operator (cf. 18a). When a pronoun is used, a covert *ruguo* is there to license the antecedent *wh*-phrase as an existential quantifier, and the pronoun is treated as an E-type pronoun (cf. 18b).

(17) a. Shei yao zhe puo-chang, wo jiu rang gei shei/ta.
who want this broken.factory I then give to who/him
'Whoever wants this broken factory, I will give it to him.'
b. Shei bu dui, wo jiu shuo shei/ta bu dui.
who not right I then say who/he not Right
'Whoever is not right, I will say he is not right.'

(18) a. NEC$_i$ [...*wh*$_i$...] [...*jiu* ...*wh*$_i$...]
b. [(*ruguo*)...$\exists x_i$...] [...*jiu* ...*pronoun*$_i$...]

C&H further clarify that in a *ruguo*-conditional, the conditional morpheme *ruguo* may, but does not have to, license a *wh*-phrase as an existential quantifier. When it does, the sentence is a *ruguo*-conditional. When it doesn't, the sentence is a bare conditional, with the two *wh*-phrases functioning as unselectively bound variables even in the presence of *ruguo*. Thus, a bare conditional is not necessarily bare in the sense that it does not have *jiu* or *ruguo*, but simply is one in which the *wh*-phrase in the antecedent is not licensed as an existential quantifier and both *wh*-phrases are bound variables (cf. 20a). Under this analysis, our earlier examples of *ruguo*-conditionals (13), repeated below as (19), can have either (20a) or (20b) as their interpretations. Therefore, they argue that there is in fact no alternation in (17) or (19), but simply an ambiguity of analysis, and that their original theory can be maintained.

(19) a. Ruguo shei yao zhe po-chang, jiu rang ta/shei
if who want this broken.factory then let him/who
dao bangongshi lai zhao wo.
to office come find me
'Whoever wants this broken factory, let him come to my office to see me.'

b. Xiangshan meiyou liang-pian xiangtong de hongye, ruguo
Xiangshan not.have two-CL same MOD red.leave if
shei zhaodao-le ta/shei jiu shi zui xingfu de ren.
who find-PERF he/who then be most happy MOD person
'There are no two identical maple leaves in Xiangshan. If anyone/someone finds them, then he will be the happiest person.'

(20) a. NEC$_i$ [*ruguo…wh*$_i$*…*] [*…wh*$_i$*…*]
b. [*ruguo*…$\exists x_i$…] [*…pronoun*$_i$*…*]

C&H's reasoning above is based on an assumption that a *wh*-phrase in the consequent clause of a conditional can only be interpreted as a bound variable, and a pronoun in a similar environment can only be an E-type pronoun. This seems to be an ad hoc assumption specifically made to accommodate the exceptions to their proposed paradigms. There is no a priori reason to exclude the E-type interpretation for *wh*-phrases and the bound variable interpretation for pronouns. If one follows C&H's argument above, then all that determines what interpretation strategy to be adopted for the anaphoric element is its form: whenever a conditional has a pair of *wh*-phrases, they are unselectively bound variables, and whenever it has a pronoun in the consequent, it is an E-type pronoun. For this reason, C&H change the term 'bare conditionals' to '*wh-wh* conditionals', and the term '*ruguo/dou*-conditionals' to '*wh-pronoun* conditionals'. A point to note, however, is that C&H (2020) have changed from their original position of bare conditionals vs. *ruguo/dou*-conditionals to *wh-wh* conditionals vs. *wh-pronoun* conditionals, which suggests that they have abandoned their claim concerning the complementary distribution of *wh*-phrases and pronouns in the consequent clause of the so-called two types of conditionals, while they still hold that the correspondence exists between sentence types and interpretation strategies, though under the new dichotomy: *wh-wh* conditionals vs. *wh*-pronoun conditionals.

Perhaps the only evidence C&H gives for their presumption above is that in a bare conditional with two *wh*-phrases, the one in the antecedent cannot be preceded by the existential verb *you* 'have', and the *wh*-phrase in the antecedent of a *ruguo*-conditional can be prefixed by *you*, as shown in (21). The contrast suggests that the *wh*-phrase in a bare conditional cannot be an existential quantifier, while the *wh*-phrase in the antecedent of a *ruguo*-conditional must be a true existential quantifier, with the anaphoric pronoun in its consequent functioning as an E-type pronoun.

(21) a. ***You** shei xian lai, shei xian chi.
have who first come who first eat
'If X comes first, X eats first.'
b. Ruguo **you** shei qiao men, ni jiu jiao ta jinlai.
If have who knock door you then ask him enter
'If someone knocks on the door, you will ask him to come in.'
(C&H 2020, p.142)

The contrast in (21) does not seem to be robust enough to draw any conclusive generalization. One can in fact find many examples with a pair of *wh*-phrases but having *you* 'have' precede the first *wh*-phrase, as in (22). If the existential verb *you* always marks the following *wh*-phrase as an existential quantifier, the *wh*-phrase in the consequent would have to be analyzed as an E-type pronoun, contra the claim of C&H.

(22) a. You shei xiang qu, shei jiu zai zheli qian ming.
have who want go who then at here sign name
'Whoever wants to go please sign your name here.'
b. Ruguoyou shei neng zai shi fenzhen nei wancheng renwu,
If have who can at ten minutes in finish task
shei jiu neng dedao zhe-ge liwu.
who then can receive this-CL present
'If anyone can finish this task in ten minutes, then he can get this present.'

Moreover, we noted in Sect. 2 that a pronoun can replace a *wh*-phrase in the consequent of a bare conditional even in the absence of *jiu*. The relevant examples are repeated here as (23). This casts serious doubt on C&H's claim that only conditionals with *jiu* allow alternations with other anaphoric elements.

(23) a. Shei yao zhe po-chang, rang gei shei/ta hao le.
who want this broken.factory give to who/him good SFP
'Whoever wants this broken-factory, give it to him then.'
b. Shei yao zhe po-chang, rang shei/ta dao bangonshi
who want this broken.factory let who/him to office
lai zhao wo.
come find me
'Whoever wants this broken factory, let him come to my office to see me.'
c. Shei xiang qu Beijing, pro bixu dao wo zheli baodao.
who want go Beijing pro must to me here register
'Whoever wants to go to Beijing, he must register with me.'

In addition, C&H's assumption that the presence of *jiu* indicates a covert *ruguo* is dubious. *Ruguo* can in general license a *wh*-phrase as an indefinite, though *jiu* alone cannot do so, suggesting that the latter does not always collocate with *ruguo*. The sentence (24a), with the connective *jiu*, is interpreted as a question; if *ruguo* is added, the sentence turns into a statement. (24a) and (24b) are thus not equivalent in meaning, which suggests that (24a) does not have a covert *ruguo* despite the presence of *jiu*. Note also that many bare conditionals with *jiu* do not actually allow the insertion of *ruguo*. For example, if *ruguo* is added to (25a) and (26a), respectively, the sentences become ungrammatical (cf. 25b and 26b). Taken together, these examples indicate that there is no strong independent evidence that *jiu* is always accompanied by *ruguo*, either overt or covert.

(24) a. Ni kanjian-le shenme ren, Zhangsan jiu hui hen gaoxing?
You see-perf what person Zhangsan then will very happy
'Which person is such that if you see him, Zhansang will be happy?'
b. Ruguoni kanjian-le shenme ren, Zhangsan jiu hui hen gaoxing.
If you see-PERF what person Zhangsan then will very happy
'If you see someone, Zhansang will be happy.'

(25) a. Wo zhibuguo shi dangshi xiangdao-le shenme jiu jiang-le shenme.
I just be then think-PERF what then say-PERF what
'It's just that I said what I happened to think of then.'
b. *Wo zhibuguo shi dangshi **ruguo** xiangdao-le shenme jiu jiang-le shenme.

(26) a. Dan deng Li Laosan huilai shuo shi shei, jiu he ta pinming.
Only wait Li Laosan return say be who then with him fight.life
'(People) are just waiting for to come back to identify that someone (who did some bad thing), and then fight with him/her for life.'
b. *Dan deng Li Laosan huilai shuo **ruguo** shi shei, jiu he ta pinming. pinming

Another point to note is that the *wh*-phrase in the antecedent clause of a *ruguo*-conditional need not be interpreted as an existential quantifier, pace C&H, as it can be interpreted as an indefinite in the sense of Heim (1982), that is, a variable, to express a rule-like statement. In this case, the *ruguo*-conditional actually employs the unselective binding strategy, which suggests that it does not need to stick to the E-type pronoun strategy, not consistent with C&H's assumption.

3.2 Bare Conditionals Without *Jiu*

C&H (2020) maintain that only those bare conditionals with *jiu* allow the alternation between *wh*-phrases and pronouns in their consequent clause, which is compatible with their theory. Therefore, they consider examples without jiu that still allow *wh*-pronoun alternations as a real challenge to their proposal. To accommodate these examples, C&H argue that they are actually composed of two independent sentences in a sequence, i.e., a rhetorical question followed by a suggestion/imperative. For them the sentences in (27) can be paraphrased as in (28).

(27) a. Shei yao zhe po-chang, rang gei ta hao le.
who want this broken.factory give to him good SFP
'Whoever wants this broken-factory, give it to him then.'

b. Shei yao zhe po-chang, rang ta dao bangonshi
who want this broken.factory let him to office
lai zhao wo.
come find me
'Whoever wants this broken factory, let him come to my office to see me.'

c. Shei xiang qu Beijing, pro bixu dao wo zheli baodao.
who want go Beijing pro must to me here register
'Whoever wants to go to Beijing, he must register with me.'

(28) a. Who wants this broken factory? Let's give it to him/her!
b. Who wants this broken factory? Let him/her come to my office to find me!
c. Who wants to go to Beijing? [You] must register with me!
(C&H 2020, p.179)

C&H's evidence for the analysis in (28) is based on intonational differences between sentences in (27) and true bare conditionals. They note that the sentences in (27) can be pronounced with a pause or a pause particle 'a' between the two component clauses. Besides, they observe that the sentences in (27) with a pronoun in their consequent differ from their counterparts with an anaphoric *wh*-phrase, as in (29), in that those in (29) must be pronounced as a compact unit with no pause or pause particle. C&H thus attribute to (29a) a structure like (30), in which the bare conditional as a whole functions as a sentential subject.

(29) a. Shei yao zhe po-chang, rang gei shei hao le.
who want this broken.factory give to who good SFP
'Whoever wants this broken-factory, give it to him then.'

b. Shei yao zhe po-chang, rang shei dao bangonshi
who want this broken.factory let who to office
lai zhao wo.
come find me
'Whoever wants this broken factory, let him/her come to my office to see me.'

(30) [[$_{\text{subject}}$ shei yao zhe po-chang, rang gei shei] [$_{\text{predicate}}$ hao le]]

The problem of this argument is three-fold. First, determining the sentence structure based on intonation is not reliable. Although it is true that the sentences in (27) can be uttered with a pause in-between, a true bare conditional (or *wh-wh* conditional, in C&H's (2020) term) can also be read with a pause, or even with a pause particle, as demonstrated in (31). In fact, this applies to many sentences presented so far, and as the sentence gets longer, it is quite natural to have a pause or pause particle between component clauses. Hence, whether there is a pause or not is not a good indicator of the interpretation strategy of the sentence in question. Second, even if one adopts the structure in (30), that does not exclude the E-type strategy, since it can yield the same truth conditions.

(31) Shei xian dao jiaoshi **ne,** wo jiu jiangli shei.
who first arrive classroom PRT I then reward who
'I will give a reward to whoever arrives at the classroom first.'

Finally, although all the counterexamples provided by P&J seem to have an imperative flavor, which motivates C&H to provide the analysis in (28), examples which are not imperatives can be easily found, as shown in (32). The second clauses of these sentences are not likely to be taken as imperatives, because their subjects are not the addressee. Note that these sentences do not have the connective *jiu* either, thus providing further evidence against the claim that *jiu* is required for the use of pronouns.

(32) a. Shei xiang na diyiming, na ta bixu fuchu hen da de nuli.
who want get top then he must give very great MOD effort
'Whoever wants to become the top student must spend great effort.'
b. Mei ci laobanxuanzhong na-ge ren, ta zhun neng chenggong.
every time boss choose which-CL person he surely can succeed
'Every time whoever was chosen by the boss could surely succeed.'
c. Jintianshei qing de ke, mingtian hai dei ta qing.
today who invite DE guest tomorrow again must he invite
'Whoever treats today has to treats again tomorrow.'

3.3 *Dou*-conditionals

Recall that contrary to C&H's claim, *dou*-conditionals in fact allow *wh*/pronouns alternation in the consequent clause. The relevant examples are repeated below (cf. (14)).

(33) a. Amei shuo: "Gei shei kan, **ta/shei** dou hui shuo wo shi haoxinhaoyi."
Amei say give who look he/who all will say I be good.will
'Amei said: 'Whomever you give to look at (it), he will say that I meant well.'
b. Zhang Manyu ai-shang shei, **ta/shei** dou hui xinxiruokuang.
Maggie.Cheung love who he/who all will extremely.happy
'Whoever Maggie Cheung falls in love with will be extremely happy.'
c. Shuo gei shei ting, **ta/shei** dou hui xiao-diao ya.
say to who liesten he/who all will laugh.fall teeth
'Whoever you tell this to, he will laugh his head off.'

C&H analyze the *wh*-phrase-containing antecedent clause in a *dou*-conditional as question-denoting, with *dou* quantifying over its set of possible answers. P&J in fact do not deny this possibility, as it is true that any plurality to the left of *dou* can be quantified over by it. What they argue for instead is that the analysis in which the universal quantifier *dou* unselectively binds two individual variables, which yields the same truth conditions, cannot be excluded.

C&H (2020) claim that in examples like (33), which are problematic for their theory, the second *wh*-phrase has a (free choice) universal reading contributed by *dou* 'all'. According to them, (33a) is interpreted as "No matter who you show (it) to, everyone will say that I meant well", with "everyone" including the one you show (it) to. The interpretation 'for all x, if letting x look at it, x will say I meant well', is an implication of the universal reading. Suppose there are three individuals in the domain of discourse: John, Tom, and Mary. The sentence (33a) under C&H's analysis would mean "if you show it to John, all the three people in question will say I meant well; and if you show it to Tom, all the three people in question will say I meant well; and if you show it to Mary, all the three people in question will say I meant well." Although this reading may well be available, the sentence also has a reading "for any one of the three people, if you show it to him, the person you show it to (not all three) will say I meant well." This reading is a real interpretation of the sentence, not just an implication of the universal (or free choice) interpretation of the second *wh*-phrase.

C&H's interpretation of (33a) entails that everyone would say I meant well once a single person among the three have looked at it. That this is not the only interpretation of (33a) can be clearly seen with the continuation in (34), where it is explicitly denied that all people will say I meant well when not all (but only some) people have looked at it, as the sentence does not sound contradictory. In fact, the second *wh*-phrase is interchangeable with the singular pronoun *ta*, as demonstrated in (33). The fact that the singular pronoun cannot be quantified over by *dou* further suggests that it is incorrect to say that the anaphoric reading of the second *wh*-phrase is an implication of the universal force of the second *wh*-phrase quantified over by *dou*.

(34) Gei Shei kan, shei dou hui shuo wo shi haoxinhaoyi.
give Who look who all will say I be good.will
Mei kan-guo de ren jiu bu hui zheme shuo.
not look-EXP MOD person then not will so Say
'Whomever you give to look at (it), he will say that I meant well. Those who haven't looked wouldn't say so.'

A similar observation can be made about our earlier examples in (14) and (15), some of which are repeated below. (35) admits two readings. On one reading, everyone should put on their clothes if any single person gets up. On the other reading, only the person who gets up needs to put on his/her clothes. With the second reading, the sentence is true if someone gets up but not all people put on their clothes. C&H's analysis only allows the first reading of (35a), which will be false in this scenario, a clearly wrong prediction. The co-varying interpretation of the two *wh*-phrases is even more prominent for (36), which can be derived by having *dou* simultaneously binds the two *wh*-variables, suggesting that unselective binding is also available to *dou*-conditionals.

(35) Ruguo shei qichuanggan shenme, biru dao shui he, na
if who get.up do what for.examplepour waterdrinktake

yan, qu maojin, shei dou yao chuan-shang yifu.

cigarette take towel who all should wear.up clothes

(i) 'If anyone gets up from bed to do something, like adding some water to drink, getting some cigarettes, or getting the towel, that person must put on his clothes.'

(ii) 'If anyone gets up from bed to do something, like adding some water to drink, getting some cigarettes, or getting the towel, everyone must put on his clothes.'

(36) Zhang Manyu ai-shang shei, shei dou hui xinxiruokuang.
Maggie.Cheung love who who all will extremely.happy

(i) 'If Maggie Cheung falls in love with anyone, that person will be extremely happy.'

(ii) 'If Maggie Cheung falls in love with anyone, everyone will be extremely happy.'

Based on our discussion in this section, we conclude that C&H's (2020) responses to the challenges raised by P&J do not stand up to scrutiny, and that their proposal including the revised one with the new dichotomy of *wh-wh* conditionals vs. *wh*-pronoun conditionals remains problematic, as we have shown that both interpretation strategies are available to both types of conditionals, which suggests that the correspondence does not exist between sentence types and interpretation strategies, pace C&H.

4 The Bound Variable Hierarchy

In consideration of the various observations discussed in the last two sections, P&J argue that there is no strict correlation between sentence types and interpretation strategies, as claimed by C&H. Both unselective binding and the E-type strategy can be applied to different types of conditionals. Besides, *wh*-phrases can take on an E-type interpretation, while pronouns can also function as bound variables. They further argue that, although *wh*-phrases and pronouns have interpretative flexibility, they have their default interpretations: *wh*-phrases tend to function as variables, and pronouns prefer the E-type strategy in Chinese. Since bare conditionals prefer unselective binding and other types of conditionals prefer the E-type strategy, it is more natural for *wh*-phrases to be used in bare conditionals, and for pronouns to be used in *dou*/*ruguo*-conditionals.[5]

[5] A reviewer asks why the default pattern should be the default pattern, but not the other way round. We suspect that this has something to do with the function of the conditional morpheme *ruguo* and the universal quantifier *dou*. In the absence of such elements, a *wh*-phrase in the

4.1 One-Case Conditionals vs. Multi-case Conditionals

Kadmon (1987, 1990) observes that English conditionals can have two interpretations. The conditional (37a) has a multi-case interpretation: it states that Sally is pleased about each case/instance of a man walking in and sitting. In contrast, the prominent reading of (37b) does not involve quantification over cases, but simply states that if the antecedent is a fact, the consequent is also a fact. Kadmon argues that English conditionals are in principle ambiguous between these two interpretations, although one of them may be more prominent for a particular conditional sentence.

(37) a. If a man walks in and he sits down, Sally is pleased.
b. If it is true that a man walked in and sits down, then Sally will be pleased.
(Kadmon 1987, p.223)

P&J note that both Chinese bare conditionals and *ruguo*-conditionals can have two interpretations as well. Under one interpretation, the sentence talks about a general situation (like a generic sentence), which is similar to the multi-case interpretation of English conditionals; under the other reading, the sentence is concerned with a particular situation or individual, much like the one-case interpretation.[6] They argue that the generic reading is derived by unselective binding, and the specific reading is generated by the E-type strategy. For example, the sentence (38a) is likely to be interpreted generically, talking about a general rule or regulation. In contrast, (38b) talks about a particular individual who bought the wrong cake.

(38) a. Shei yao zhe puo-chang, wo jiu rang gei shei.
who want this broken.factory I then give to who
'Whoever wants this broken factory, I will give it to him.'
b. Zuotian shei mai-cuo-le dangao wo jiu fa shei
yesterday who buy-wrong-PERF case I then fine who
de qian.
POSS money
'Who bought the wrong cake yesterday, I will fine him.'

The multi-case reading is the prominent reading of the classic donkey sentences such as *If a famer owns a donkey, he beats it*, which asserts that for every case of a farmer

antecedent clause takes on its default interpretation, i.e., a free variable, and the anaphoric *wh*-phrase has to be a variable too. The conditional morpheme *ruguo* and the universal quantifier *dou* have the preferred option to license the *wh*-phrase in the antecedent clause as an existential quantifier, and the anaphoric element has to be an E-type pronoun then. In this light, *ruguo* seems to differ from English *if*. The semantic contribution of an *if*-clause can be analyzed as providing the restriction for a (overt or covert) quantifier (Lewis 1975, Kratzer 1986), and the morpheme *if* is semantically vacuous. But *ruguo* has an obvious role to play for determining the interpretation of the *wh*-phrase in the antecedent clause.

[6] Lin (1996) also observes that Chinese donkey conditionals have a multi-case or one-case interpretation, but he differs from P&J (and this paper) in that he claims that pronouns are incompatible with multi-case bare conditionals.

owning a donkey, he beats it. The semantics of such sentences can be given using unselective binding.[7] Take the multi-case donkey conditional (38a) for an illustration. It can be analyzed as having the implicit necessity operator unselectively bind the world variable and the individual variable, as in (39a).[8] On the other hand, the one-case interpretation of (38b) can be derived by using the E-type strategy, as in (39b), where the necessity operator binds only the world variable, so that the sentence does not involve multiple instances of someone buying a wrong cake (in a world). Similarly, a *ruguo*-conditional conditional admits two readings as well, as shown in (40), where (40a) corresponds to the multi-case reading, and (40b), the one-case reading.

(39) a. $NEC_{w,x}$[x wants this broken factory in w][I'll give it to x in w]
     b. NEC_{w}[∃x[x bought the wrong cake in w]][I'll fine **that person** in w]

(40) Xiangshan meiyou liang-pian xiangtong de hongye, ruguo shei zhaodao-le,
     Xiangshan not.have two-CL same MOD red.leave if who find-PERF
     shei jiu shi zui xingfu de ren.
     who then be most happy MOD person
     (i) $NEC_{w,x}$[…x…in w][…x…in w]:
         Anyone who finds them will be the happiest person.
     (ii) NEC_{w}[∃x…in w][…he…in w]:
         If someone has found them, he will be the happiest person.

For *dou*-conditionals, P&J maintain that simply treating the *wh*-phrase as a variable bound by *dou* 'all' yields the correct interpretation in (41a), which would be indistinguishable from treating the *wh*-phrase-containing clause as a question in which the *wh*-phrase functions as an existential quantifier, and the pronoun is treated as an E-type pronoun, shown in (41b).

(41) Ni jiao shei jinlai, wo dou jian ta.
     you ask who enter I all see him
     'Whoever you ask to come in, I'll see him.'
     (a) ∀x[[person(x) and you ask x to come in] → I will see x]
     (b) ∀p[[∃x[person(x) ∧ p = you ask x to come in]] → I'll see the person you ask in the event of p]

[7] For multi-case conditionals, Heim (1990) argues that a DRT-style unselective binding analysis and the alternative situation-based E-type analysis is semantically equivalent. (38a) thus can also be analyzed as quantification over minimal situations: each minimal situation s that contains a person who wants this broken factory can be extended to a situation s' in which I will give the factory to that person in s. To the extent that the unselective binding analysis and the E-type analysis handle multi-case conditionals equally well, there is no ground to exclude either one, and this provides further support to our null hypothesis that both unselective binding and the E-type strategies are possible.

[8] We remain neutral as to the nature of the implicit operator in donkey conditional sentences (e.g., whether it is an implicit modal or adverb of quantification), and simply follow Heim (1982) to posit a necessity operator binding both the world variable and the individual variable.

P&J summarize the idea above as the bound variable hierarchy in (42). The anaphoric elements on the left are more suitable for being used as bound variables,[9] and those on the right is less suitable for being so used; on the other hand, the one on the right is more suitable to be an E-type pronoun,[10] and the one on the left is less so. Since the hierarchy is taken to be the default pattern, exceptions to it are expected, though they need additional contextual support.

(42) Bound Variable Hierarchy[11]
wh-phrases/reflexives >> pronouns/demonstratives

[9] As the hierarchy makes it clear, *wh*-phrases in Mandarin are able to function as variables. A reviewer raises a question concerning the status of *wh*-phrases as variables: what excludes the *wh*-phrase from being bound by the subject quantifier in (i) if it can be a variable? Previous works (e.g., Huang 1982, Cheng 1991, Li 1992, Lin 1996) have shown that the non-interrogative use of *wh*-phrases, especially their use as indefinite NPs, is subject to certain licensing requirements. *Wh*-indefinites are like polarity items and typically occur in downward entailing environments. Therefore, it is not the case that they can freely occur. The reviewer also wonders why the bound variable reading of *wh*-phrases only appears in the conditional type of donkey sentences instead of the relative clause type such as (ii). In a similar vein, the indefinite *wh*-phrase cannot be licensed in this case. In fact, if one follows Tsai's (1994, 1999) proposal of *wh*-contruals, even the interrogative *wh*-phrase is semantically a variable subject to binding from the question operator directly merged in CP. In general, it seems that *wh*-phrases as variables must be bound by A'-operators instead of nominal quantifiers.

(i) *Mei-ge ren dou zhi xihuan shei de zuoping.
every-CL person all only like who POSS work
Every person only likes his own work.

(ii) *Mei-ge gen shei shuo-guo hua de dou yao ba shei de mingzi
every-CL person who say-EXP word MOD all will BA who POSS name
xie xialai.
write down
Every person who has talked to someone will write down his name.

[10] A reviewer notes that Mandarin *wh*-phrases do not behave like pronouns outside bare conditionals, as shown in (iii) and (iv), and doubts whether they have E-type interpretations. This is in fact a welcome result for the bound variable hierarchy, since it dictates that *wh*-phrases are less suitable and thus quite restricted for being used as E-type pronouns. A question worth pursuing is how E-type *wh*-phrases are restricted. One may resort to a competition mechanism and say that whenever a pronoun is available, a *wh*-phrase cannot be used as an E-type pronoun.

(iii) #Yi-ge nongmin zou-le jinlai. Shei zou-le xialai, dian-le yi
one-CL farmer walk-PERF enter who sit-PERF down order-PERF one
bei pijiu.
glass beer
A farmer came in. He sat down, and ordered a beer.

(iv) #Ruguo you ren yao zhe po-chang, jiu rang shei lai
one-CL have person want this broken.factory then let who come
zhao wo.
find me
'Whoever wants this broken factory, let him/her come to to see me.'

[11] We thank a reviewer for raising a question about the nature and status of this hierarchy in natural languages. Indeed, the precise nature of this proposed hierarchy and its empirical consequences remain to be further investigated, but it seems necessary to restrict its application to languages in which wh-phrases have the option to be used as indefinites; otherwise, languages like English

4.2 Accounting for *Wh*/pronoun Alternations

It has been observed that pronouns in Mandarin differ from their English counterparts in that they do not easily function as bound variables, even when c-commanded by quantificational NP antecedents, and the bound variable use of pronouns is subject to some syntactic constraint (cf. Huang 1987; Aoun and Li 1990, 1996).[12] Thus, it is justified to say that pronouns rank lower in the bound variable hierarchy, and that they are preferably used as unbound or referential pronouns.

Apart from this hierarchy, the well-formedness of a donkey sentence in Chinese is also affected by discourse coherence requirement. The idea is that the two clauses of a conditional sentence need to have some sort of connection to be interpreted as a coherent mini-discourse; otherwise, the sentence can be understood to consist of two independent sentences, resulting in a degraded status for being a conditional. A primitive bare conditional like (1a), repeated as (43) below, does not allow the *wh*-phrase in the consequent clause to alternate with a pronoun. The unacceptability of the pronoun can be understood in the following way (see also P&J). Since the bare conditional does not have any marker indicating the connection between the two clauses and a pronoun is more likely to be used freely, the two clauses thus have the preferable option of being understood as two independent sentences; therefore, the whole discourse consisting of the two clauses will not be considered as a conditional. When a sentence connective like *jiu* is added, the connection between the two clauses is then established and a pronoun becomes much better in the consequent. The conditional morpheme *ruguo*/*yaoshi* 'if' by default licenses a following *wh*-phrase as an existential quantifier, and consequently a pronoun is the preferred option.

(43) Shei xian lai, shei/*ta xian chi.
who first come who/he first eat
'If X comes first, X eats first.'

In connection with discourse coherence, it should be noted that it is not the case that any two *wh*-clauses can be conjoined to form a bare conditional. The examples in (44) are unacceptable. The intuition is that in both examples the two component clauses lack connection so that they are likely to be interpreted as two independent sentences. If a sentence connective *jiu* 'then' is added to link the two clauses, these examples become grammatical (cf. 45).

in which wh-phrases are more operator-like would be predicted to have bare conditionals as well.

[12] Native speaker's judgments on whether Chinese pronouns can function as bound variables when c-commanded by quantificational NPs seem to show some variation, though, in general, the bound variable use of Chinese pronouns do not sound as natural as their English counterparts.

(44) a. *Shei qu-le Beijing, shei qu-le Shanghai.
who go-PERF Beijing who go-PERF Shanghai
'Whoever went to Beijing also went to Shanghai.'
b. *Shei jie-le wo de shu, shei jide huan gei wo.
who borrow-PERF I poss book who remember return to me
'Whoever has borrowed books from me should remember to return them to me.'

(45) a. Shei qu-le Beijing, shei jiu qu-le Shanghai.
who go-PERF Beijing who then go-PERF Shanghai
'Whoever went to Beijing also went to Shanghai.'
b. Shei jie-le wo de shu, shei jiu jide huan gei wo.
who borrow-PERF I POSS book who then remember return to Me
'Whoever has borrowed books from me should remember to return them to me.'

4.3 Other Types of Anaphoric Elements

Chinese seems to make special use of a pair of identical NPs to express a general rule or regulation. In addition to pairs of *wh*-phrases, one can also resort to pairs of indefinite NPs (46) or reflexives (47) (cf. Wen 2006, Liu 2021).

(46) a. Yi ren zuo shi, yi ren dang.
one person do thing one person be.responsible
'For every person x, if x has done something wrong, x should be responsible for it.'
b. Yi ren jiang hua, yi ren fuze.
one person say word which-CL person take.responsibility
'For every person x, if x said something, x should be responsible for what he said.'
c. Ta zongshi jian yi-ge ren, ai yi-ge ren.
he always see one-CL person love one-CL person
'For every person x, if he sees x, he loves x.'

(47) Ziji zuo cuo shi, ziji chengdan.
self do wrong thing self take.responsibility
'For every x, if x has done something wrong, x must be responsible for it.'

Note that sentences in (46) and (47) have a very similar meaning to their bare conditional counterparts, with indefinite NPs or reflexives replaced by *wh*-phrases, as in (48).

(48) a. Shei jiang hua, shei fuze.
who say word who be.responsible
'For every person x, if x said something, x should be responsible for what he said.'

b. Ta zongshi jian shei ai shei.
he always see who love who
'For every person x, if he sees x, he loves x.'

c. Shei zuo cuo shi, shei chengdan.
who do wrong thing who take.responsibility
'For every x, if x has done something wrong, x must be responsible for it.'

Universal NPs like *mei-ge ren* 'everyone' are also able to occur in pairs to convey a meaning that is typical of bare conditionals, as shown in (49). Li and Pan (2022) argue that the second universal NPs in the relevant sentences function as bound variables.

(49) a. Mei-ge guojia you mei-ge guojia de shiji qingkuang.
every-CL country have every-CL country POSS actual situation
'For every country x, x has x own actual situation.'

b. Mei-ge xuesheng zuo hao mei-ge xuesheng yao zuo de shi.
every-CL student do good every-CL student want do REL thing
'For every student x, x should do well the things he needs to do.'

If one takes the *wh-wh* conditionals in (48) and those sentences in (46), (47), and (49) to be cases that instantiate the same general strategy, i.e., by using pairs of identical NPs to convey a sort of co-varying meaning, we have an answer to the question concerning the occurrence of *wh*-indefinites in the consequent clauses of conditionals (cf. Chierchia 2000, Liu 2017). The unselective binding account of *wh-wh* conditionals assumes that *wh*-phrases are indefinites functioning as variables, but a *wh*-phrase is licensed as an indefinite typically in downward entailing environments such as the antecedent of conditionals, instead of upward entailing environments like the consequent of conditionals. It is mysterious why a *wh*-indefinite can occur in the consequent of a conditional.

Another puzzling aspect of Chinese bare conditionals is that the *wh*-indefinite in the consequent has to be anaphoric to the *wh*-phrase in the antecedent rather than introducing a new discourse referent, which violates the novelty condition of Heim (1982). This motivates Chierchia (2000) to argue that Mandarin *wh*-phrases are indefinite pronouns, or leads to some other approaches to bare conditionals, like the question-based approach (Liu 2017). However, if one considers data beyond *wh-wh* conditionals, such as sentences in (46), where an indefinite *yi-ge NP* 'one NP' appears in the antecedent as well as in the consequent, the occurrence of *wh*-indefinites in the consequent no longer seems to be that mysterious. No one would doubt that NPs like *yi-ge ren* 'one person' are genuine indefinite NPs. Treating *wh-wh* conditionals as a different type of sentence than those in (46) and (47) would miss a more general picture: Chinese seems to have a special device of using pairs of identical NPs to express a co-varying meaning, each of these NPs semantically functioning as bound variables. Of course, not all sentences with two identical NPs are well-formed under such a use, but the strategy is quite productive. Looked at in this way, the sentences with two identical NPs can be subsumed under the same sentence type as *wh-wh* conditionals, all of which use unselective binding to achieve the co-varying interpretation.

It is also possible to use two pronouns in a bare conditional, although they have to be null pronouns, as demonstrated by the contrast in (50). C&H (2020) challenge P&J by arguing that (50b) should be grammatical if pronouns can function as bound variables in conditionals. But in fact (50b) is not problematic for P&J, because as the bound variable hierarchy dictates, the overt pronoun *ta* 'he' is by default interpreted as a free pronoun, and therefore cannot be used in bare conditionals if no other contextual support is provided. Moreover, it is stipulated in DRT that pronouns (and other definite NPs) are subject to a familiarity condition, and as such the first pronoun in (50b) cannot introduce a discourse referent. On the other hand, null pronouns are perfectly suited to be variables. The contrast in (50) is paralleled by (51), where the overt pronoun cannot function as a variable bound by the quantificational NP antecedent, though the null pronoun can happily do so.

(50) a. pro yao xiang fu, pro xian xiu lu.
pro will want rich pro first build road
'For every x, if x wants to get rich, x needs to build roads first.'
b. *Ta yao xiang fu, ta xian xiu lu.
he will want rich he first build road
'For every x, if x wants to get rich, x needs to build roads first.'

(51) a. ??Mei-ge ren dou shuo ta xihuan Xiaoming.
every-CL person all say he like Xiaoming
'For every person x, x says x likes Xiaoming.'
b. Mei-ge ren dou shuo pro xihuan Xiaoming.
every-CL person all say pro like Xiaoming
'For every person x, x says x likes Xiaoming.'

How could one make sense of the use of identical NPs? The essence of the DRT account of anaphora is to model it as the co-variation of occurrences of the same variable through variable binding. The discourse referent introduced by the antecedent NP and the anaphoric element are translated as the same variable bound by the same quantifier, and the semantic interpretation rule ensures that they are assigned values by the same assignment function, and thus always co-vary. Identical NPs may be seen as a morphological marking of occurrences of the same variable, and in a sense, Chinese is more logically transparent in that it wears semantic co-variation on its sleeve.[13]

[13] That Chinese is more logically transparent seems to be a general property of the language, and may also be reflected by another fact, namely that it obeys a Scope Isomorphism Principle (Huang 1982, Lee 1986), which states that the surface c-command relation between logical operators determines their scope relation.

5 Conclusion

In this paper, we first reviewed C&H's (1996) treatment of Chinese donkey sentences, and discussed P&J's (1997, 2015) objections to the alleged complementary distribution between *wh*-phrases and pronouns in the consequent clause of donkey conditionals and the claim that the two types of conditionals are associated with their own interpretation strategies. We then focused on C&H's (2020) defense of their 1996 proposal, in which they brought up a new dichotomy of *wh-wh* conditionals vs. *wh*-pronoun conditionals, replacing their original dichotomy of bare conditionals vs. *ruguo/dou* conditionals and abandoning the complementary distribution between *wh*-phrases and pronouns in the relevant consequent classes, though they still maintain that the correspondence exists between sentence types and interpretation strategies. We have shown that C&H (2020) have not really dismissed P&J's objections, and their problems remain, that is, the correspondence does not exist between sentence types and interpretation strategies. The bound variable hierarchy, first proposed in P&J, was employed to accommodate the problematic data, and the discourse coherence relation helps explain why primitive bare conditionals are not compatible with the pronoun *ta*. The take-home message is that the null hypothesis of P&J can be maintained: that is, both *wh*-phrases and pronouns can appear in the consequent clause of bare conditionals and *ruguo/dou*-conditionals, functioning as bound variables and E-type pronouns, and both interpretation strategies are employed by the two types of conditionals, though they have their default interpretation strategy and the non-preferred strategy requires additional contextual support.

References

Aoun, J., Li, Y.-H: Minimal disjointness. Linguistics **28**, 189–203 (1990)

Aoun, J., Li, Y.-H: Two cases of logical relations: Bound pronouns and anaphoric relations. Linguistics 28, 189–203. In: Freidin, R. (ed.) Current Issues in Comparative Grammar, pp. 346–373. Springer, Dordrecht (1996)

Chen, S.陳舜婷: Anaphora and sentence patterns in *if-who* conditionals [如果"誰"條件句的照應和句式研究]. J. Language Literat. Stud. [語文學刊] **12**,10–11 (2015).

Chen, Y.: Deriving *wh*-correlatives in Mandarin Chinese: *Wh*-movement and (island) identity. In: NELS 50 Proceedings (2019)

Cheng, L.-S.: On the typology of *wh*-questions. Ph.D. dissertation, MIT (1991)

Cheng, L.-S., Huang C.-T.: Two types of donkey sentences. Natural Lang. Semant. **4**, 121–163 (1996)

Cheng, L.-S., Huang, C.-T.: Revisiting donkey anaphora in Mandarin Chinese: a reply to Pan and Jiang (2015). Int. J. Chinese Linguist. **7**(2), 167–186 (2020)

Chierchia, G.: Chinese conditionals and the theory of conditionals. J. East Asian Linguis. **9**(1), 1–54 (2000)

Cooper, R.: The interpretation of pronouns. In: Heny, F., Schnelle, H. (eds.) Syntax and Semantics 10: Selections from the third Gröningen Round Table, pp. 61–92. Academic Press, New York (1979)

Evans, G.: Pronouns. Linguistic Inquiry **11**, 337–362 (1980)

Heim, I.: The semantics of definite and indefinite noun phrases. Ph.D. dissertation, University of Massachusetts, Amherst (1982)

Heim, I.: E-type pronouns and donkey anaphora. Linguist. Philos. **13**, 137–178 (1990)

Higginbotham, J.: Pronouns and bound variables. Linguistic Inquiry **11**, 679–780 (1980)

Huang, A.: A DR-theoretic account of Chinese donkey sentences. MA thesis, Hunan Normal University (2003)

Huang, C.-T.: Logical relations in Chinese and the theory of grammar. Ph.D. dissertation, MIT (1982)

Huang, C.-T.: Remarks on empty categories in Chinese. Linguistic Inquiry **18**(2), 321–337 (1987)

Huang, Y.: On the form and meaning of Chinese bare conditionals: Not just whatever. Ph.D. dissertation, University of Texas at Austin (2010)

Kadmon, N.: On unique and non-unique reference and asymmetric quantification. Ph.D. dissertation, University of Massachusetts, Amherst (1987)

Kadmon, N.: Uniqueness. Linguist. Philos. **13**, 273–324 (1990)

Kadmon, N.: Formal Pragmatics. Blackwell, Oxford (2001)

Kamp, H.: A theory of truth and semantic representation. In: Groenendijk, J., Janssen, T., Stokhof, M. (eds.) Formal methods in the study of language, pp. 277–322. Mathematical Centre Tracts, Amsterdam (1981)

Kratzer, A.: Conditionals. In: Farley, A., Farley, P., McCollough, K. (eds.) Papers from the Parasession on Pragmatics and Grammatical Theory, pp. 115-135. Chicago Linguistics Society, Chicago (1986)

Lee, H.-T.: Studies on quantification in Chinese. Ph.D. dissertation, UCLA (1986)

Li, H.: Mandarin *wh*-conditionals: a dynamic question approach. Nat. Lang. Seman. **29**, 401–451 (2021)

Li, Z. 李貞亮, Pan, H. 潘海華.: Double *Mei* constructions: bound variable interpretations and salient reading [雙"每"句中的受約變量及其優勢解讀]. J. Foreign Lang. **45**(12), 47–60 (2022)

Li, Y.-H.: Indefinite *wh* in Mandarin Chinese. J. East Asia Linguist. **1**(2), 125–155 (1992)

Lin, J.-W.: Polarity licensing and wh-phrase quantification in Chinese. PhD thesis, University of Massachusetts, Amherst (1996)

Liu, M.: Varieties of Alternatives: Focus Particles and wh-Expressions in Mandarin. Springer and Peking University, Berlin and Beijing (2017)

Liu, T. 劉探宙.: Chinese co-variation anaphora and English donkey sentences [漢語倚變同指照應格式與英語驢句]. In: Paper presented at The 8th symposium on recent advances in Chinese syntax and semantics [第八屆漢語句法語義前沿討論會] (2021)

Lewis, D.: Adverbs of quantification. In: Keenan, E. (ed.) Formal Semantics of Natural Language, pp. 3–15. Cambridge University Press, Cambridge (1975)

Luo, Q., Crain, S.: Do Chinese *wh*-conditionals have relatives in other languages? Language Linguist. **12**(4), 753–798 (2011)

Pan, H., Jiang, Y.: NP interpretation and Chinese donkey sentences. In: Proceedings of the Workshop on Interface Strategies in Chinese: Syntax and Semantics of Noun Phrases. Summer Institute of Linguistics Society of America, Cornell University (1997)

Tsai, W.-T.: On economizing the theory of A-bar dependencies. PhD dissertation, MIT (1994)

Tsai, W.-T.: On lexical courtesy. J. East Asian Linguis. **8**(1), 39–73 (1999)

Wang, G. 王廣成.: On the syntax and semantics of two types of conditionals [兩種條件句式的語法、語義解釋]. Linguistic Sciences [語言科學], 5(6), 42–51 (2006)

Wen, B. 温賓利: "Donkey sentences" in English and "*wh*…*wh*" construction in Chinese [英語驢句與漢語的"什麼…什麼"句]. Modern Foreign Languages [現代外], 3, 1-13 (1997)

Wen, B. 温賓利.: A relative analysis of Mandarin "*wh*…*wh*" construction ["什麼…什麼句": 一種關係結構]. Modern Foreign Languages [現代外語], 82, 1–17 (1998)
Wen, W. 文衛平.: Studies on donkey sentences [英漢驢子句研究]. PhD dissertation, Beijing Language and Culture University (2006)
Xiang, Y.: A hybrid categorial approach to question composition. Linguist. Philos. **44**(3), 587–647 (2020). https://doi.org/10.1007/s10988-020-09294-8

Updating Contexts: Ignorance and Concession

Yi-Hsun Chen(✉)

Graduate Program of Teaching Chinese as a Second Language, National Chengchi University, New Taipei, Taiwan
yhchen21@nccu.edu.tw

Abstract. Superlative modifiers (SMs) such as *at least* shows an intriguing ambiguity between a reading conveying speaker concession (CON) and an epistemic reading conveying speaker ignorance (EPI). While it is an going debate on whether such EPI-CON ambiguity should be a case of homophony (e.g., Nakanishi and Rullmann 2009; Cohen and Krifka 2014) or arises from only one lexical entry (Biezma 2013), it seems to receive relatively little attention on how an assertion with *at least* under the two readings updates the discourse. This paper fills the research gap by considering how an assertion with *at least* in the discourse fares with respect to diagnostics of (not-)at-issue content, and presenting a pragmatic analysis capturing their different discourse profiles under the system of conversational scoreboard in Farkas and Bruce (2010) and subsequently developed by others (e.g., Malamud and Stephenson 2015). In particular, the speaker is committed to both the truth of the prejacent and the falsity of the higher alternatives under CON. By contrast, the speaker is only committed to the *possibility* that the prejacent is true and the *possibility* that some higher alternative is true in *subsequent discourse*. This modal flavor of *at least* arises at the level of discourse, rather than lexical semantics (e.g., via a covert epistemic modal). Finally, it is shown that the pragmatic analysis can be extended to expressions in other languages such as Chinese *zhishao* 'at least', which also demonstrates the familiar EPI-CON ambiguity.

Keywords: Concession · Conversational scoreboard · Ignorance · Superlative modifiers

1 Introduction

Nakanishi and Rullmann (2009) (henceforth N&R) observe that English sentences containing *at least* can have two readings: an epistemic reading (EPI) and a concessive reading (CON), the first conveying ignorance on the part of the speaker about the actual state of affairs and the second conveying an evaluation by the speaker about a settled fact. Assuming a context in which there are three medals, the highest being gold, followed by silver and then bronze, the examples in (1) can be used to illustrate the distinction. EPI conveys that the speaker is sure that Mary won either a silver medal or a gold medal, but

D. Deng et al. (Eds.): TLLM 2022, LNCS 13524, pp. 52–75, 2023.
https://doi.org/10.1007/978-3-031-25894-7_3

cannot say anything more definitive. CON conveys that Mary's winning a silver, while less desirable than winning a gold medal, is better than her winning a bronze:[1]

(1)	a. Mary won **at least** a [silver]$_F$ medal	**EPI**
	b. (Mary didn't win a gold medal, but) **at least** she won a [silver]$_F$ medal.	**CON**

The use of *at least* signals that the prejacent is interpreted against other possible propositions and that these propositions are ranked in order of preference. N&R note that CON has a "settle-for-less" flavor in the sense that it falls short of the intended goal of winning the gold. EPI is neutral in the sense that there need not be any prior expectations about what Mary would win, or whether she would anything.

This is also in evidence in statements with numerals, where the logic of numbers determines the ranking between the relevant propositions. Assuming that we are talking about a game of dice where the minimum someone can get is 1 and the maximum is 6, (2a) is consistent with Mary getting three, four, five or six; Instead, (2b) takes it as settled that she did not get four or more, which would have been better, but indicates that she could have got one or two, which would have been worse:

(2)	a. Mary got **at least** [three]$_F$.	**EPI**
	b. **At least** Mary got [three]$_F$.	**CON**

More generally, while the source of the ranking between the prejacent and the alternatives may vary, the contribution of *at least* remains fixed. It leads to two distinct readings, EPI and CON.

It is an ongoing debate whether the two readings in question represent a case of accidental homonymy or whether there is a deeper connection between them. N&R considers the EPI-CON ambiguity as a case of homophony (see also Cohen and Krifka 2014), while Biezma (2013) suggests a unified account of the two readings. This paper remains neutral on this debate while taking a closer look at the discourse profile of these two readings of *at least* with respect to the diagnostics of (not-)at-issue content, considering how an assertion with *at least* on the two readings updates the discourse and presenting a pragmatic analysis capturing their different discourse profiles under the system of conversational scoreboard in Farkas and Bruce (2010).

The rest of this paper proceeds as follows. Section 2 is devoted to the information status of concessive *at least* and epistemic *at least*, serving as an empirical background

[1] Kay (1992: 311) distinguishes three uses of *at least* in English: a scalar use, an evaluative use and a rhetorical use. The three uses are illustrated in (i).

(i)	a. Mary received calls from **at least** three soldiers.	Scalar.
	b. **At least**, this one is cooked.	Evaluative.
	c. I see her every day, **at least** when I'm in town.	Rhetorical

Kay's scalar use corresponds to N&R's EPI and his evaluative use corresponds to N&R's CON. This paper is concerned with the discourse profile of these two uses of *at least*.

for the pragmatic analysis in terms of conversational scoreboard to be presented in Sect. 3. Section 4 briefly discusses Chinese *zhishao* 'at least', which also shows the familiar EPI-CON ambiguity. Section 5 is the conclusion.

2 The Discourse Profile of *At Least*

This section proceeds as follows. Section 2.1 introduces the distinction between at-issue and not-at-issue information content, and Sects. 2.2 # 2.3 present empirical facts that **(a)** under both readings, the information content of the prejacent is always at-issue; **(b)** under both readings, the information content of the lower alternatives is at-issue; **(c)** under both readings, the information content of the ranking between the prejacent and its alternatives seems to be not-at-issue; **(d)** under CON, the information content of the higher alternatives can be at-issue or not-at-issue; while under EPI, it is at-issue; **(e)** the speaker's discourse commitments are different under the two readings: under CON, the speaker is committed to both the truth of the prejacent and the falsity of the higher alternatives; by contrast, under EPI, the speaker is *tentatively* committed to the truth of the prejacent or the higher alternatives.

2.1 At-Issue vs. Not-At-Issue Content

Recent studies on appositves, evidentials, and expressives among other expressions, have motivate the view that it is necessary to distinguish meaning contributions further than our traditional categories of what is asserted, what is presupposed and what is implicated (e.g., Potts 2005, 2007; McCready 2008, 2010; Simons et al. 2010; Murray 2010, 2014; AnderBois et al. 2010, 2015; Tonhauser 2012; Koev 2012, 2013; Tonhauser et al. 2013; Syrett and Koev 2015; among many others). In particular, many of these analyses maintain a distinction in propositional content between what is at-issue and what is not-at-issue. For example, Simons et al. (2010) suggests the following characterization of the notion of at-issueness, within the framework of Question-Under-Discussion (QUD).

(1) The characterization of at-issueness (Simons et al. 2010: 323)

  a. A proposition p is **at-issue** iff the speaker intends to address the QUD via $?p$.
  b. An intention to address the QUD via $?p$ is **felicitous** only if:

    i. $?p$ is relevant to the QUD, and
    ii. The speaker can reasonably expect the addressee to recognize this intention.

The definition of at-issue content in (1) suggests the following distinction between at-issueness and non-at-issueness: A proposition expressed by a constituent is at-issue if it contributes to the ordinary semantics of the clause in which it is located, and entails that some possible answer to the QUD is false; otherwise the proposition is not at-issue. Along this line of thought, Tonhauser (2012) further suggests three discourse properties of at-issue content: **(a)** at-issue content can be directly assented or dissented with; **(b)** at-issue content addresses the question-under-discussion; **(c)** at-issue content determines the relevant set of alternatives. Given these discourse properties of at-issue content, the following two tests have been applied to diagnose whether a given (propositional) content is at-issue or not-at-issue in the discourse (Simons et al. 2010; AnderBois et al. 2010, 2015; Murray 2010, 2014; Tonhauser 2012; Koev 2012, 2013).

(2) Direct Response Test
Only at-issue content can be directly targeted by the addressee, e.g., by replies like "Yes", "No", "That's not true", etc., in subsequent discourse.

(3) Answerability Test
Only at-issue content can be employed by interlocutors to answer questions.

Below, I briefly illustrate how these two tests distinguish not-at-issue content from at-issue content. Let's first consider the direct response test. In (4), speaker A's utterance containing a clause-medial appositive (i.e., *who was talking to Mary a minute ago*) contributes to both at-issue meaning and not-at-issue meaning, as in (5). Crucially, at-issue meanings are directly challengeable but not-at-issue meanings are not (as indicated by speaker B''s reply). In this line, the example with expressives (i.e., *that bastard*) in (6) makes the same point: speaker A's utterance conveys a meaning that Bill is a bastard; crucially, this meaning is not directly challengeable, in contrast to the at-issue meaning that Bill stole my car.

(4) Clause-medial appositives (adapted from Koev 2012: (6))

A: John, who was talking to Mary a minute ago, has gone home.
B: No, he hasn't. He is still at the party.
B': #No, he wasn't. He was talking to Stacy.

(5) a. At-issue meaning: John has gone home.

b. Not-at-issue meaning: John was walking to Mary a minute ago.

(6) Expressives (adapted from Potts 2007: 168)

A: That bastard Bill stole my car.
B: No, he didn't. It was John that stole it.
B': #No, he isn't (a bastard). Bill is a very nice person.

(7) a. At-issue meaning: Bill stole my car.

b. Not-at-issue meaning: Bill is a bastard.

Next, let's consider the answerability test. In (8), speaker B's utterance containing a clause-medial appositive (i.e., *who had prostate cancer*) contributes to both at-issue meaning and not-at-issue meaning, as in (9). Crucially, only at-issue meaning can be used as an answer to the question-under-discussion: even though the semantic content of the appositive does answer the question, speaker B's response (in contrast to speaker B"s) is infelicitous in (8). Put differently, the test indicates that the semantic content of the appositive is not-at-issue and thus cannot be used to answer questions.[2]

(8) Clause-medial appositives (borrowed from Koev 2012: (11))

A: What disease did Tammy's husband have?
B: #Tammy's husband, who had prostate cancer, was treated at the Dominican Hospital.
B': Tammy's husband had prostate cancer.

(9) a. At-issue meaning: Tammy's husband was treated at the Dominican Hospital.

b. Not-at-issue meaning: Tammy's husband had prostate cancer.

Summing up, we have seen that **(a)** not-at-issue content cannot be directly assented or dissented with, in contrast to at-issue content; **(b)** not-at-issue content cannot be used to answer the QUDs, contrasting with at-issue content. Although at-issue content is roughly what is proffered (contributed by a regular factual assertion), the group of expressions contributing to not-at-issue content is heterogeneous: it includes appositives, expressives, and among other expressions. In the next two subsections, we discuss how concessive *at least* and epistemic *at least* fare with the two tests concerning the at-issueness vs. not-at-issueness.

2.2 The Information Status of Concessive *At Least*

First of all, let us observe that under concessive *at least*, the propositional content of the prejacent and that of its higher alternatives are both at-issue. Suppose that Adam, Bill and Chris are the three relevant individuals in the discourse. Consider (10).

(10) A: Who did John invite? Did John invite Adam, Bill and Chris?

B: No, John didn't invite Adam, Bill and Chris/everyone.
But, **at least** he invited [Adam and Bill]$_F$.

In (10), speaker A explicitly raises an issue concerning whether the content of the higher alternative (i.e., John invited Adam, Bill and Chris) is true and speaker B's assertion serves as a proposal to settle the raised issue. These discourse moves indicate that the content of the higher alternative is at-issue (by the answerability test). In the same

[2] Although the illustrated examples for the two diagnoses involve only expressives and appositives, expressions contributing to not-a-issue content go far beyond those two types; see Murray (2010, 2014) on evidentials, and Simons et al. (2010), Tonhauser (2012), Tonhauser et al.(2013) on how and why projective content projects in general and how the projection behavior is connected with information structure.

vein, the three possible responses (to speaker B's assertion) in (11) # (13) further illustrate that not only the propositional content of the higher alternative, but also that of the prejacent are at-issue (by the direct response test).

(11) C: That's true!
(12) C: No, that's not true! John only invited Adam.
(13) C: No, that's not true! John invited all of them.

In (11), speaker C **assents** to speaker B's assertion with respect to the content that the higher alternative (i.e., John invited Adam, Bill and Chris) is false and that the prejacent (i.e., John invited Adam and Bill) is true. This indicates that the propositional content of the prejacent and that of its higher alternative are both at-issue. In (12), speaker C **assents** to speaker B regarding the content that the higher alternative (i.e., John invited Adam, Bill and Chris) is false, but **dissents** with speaker B from the content that the prejacent is true. This again indicates that the propositional content of the prejacent and that of its higher alternative are both at-issue. Crucially, notice that the lower alternative (i.e., John only invited Adam) is involved as part of speaker C's response addressing speaker A's question. This indicates that the propositional content of the lower alternative is also at-issue. Finally, in (13), speaker C **dissents** with speaker B on the content that the higher alternative is false. This indicates that the propositional content of the higher alternative is at-issue.[3]

Next, let us observe that while the propositional content of the prejacent remains at-issue, the propositional content of the h

(14) A: Who did John invite for the party? Did he invite [Adam and Bill]$_F$?

B: B: Yeah..., **at least** he invited [Adam and Bill]$_F$.
C: **Hey, wait a minute!** You mean he didn't invite (even) Chris?.

igher alternative can be not-at-issue. Suppose that Adam, Bill and Chris are the three relevant individuals. Consider (14).

In (14), speaker A explicitly raises an issue concerning whether the content of the prejacent (i.e., John invited Adam and Bill) is true and speaker B's assertion serves as a proposal to settle the raised issue. These discourse moves indicate that the content of the prejacent is at-issue (by the answerability test). Crucially, speaker C is entitled to use the phrase "hey, wait a minute" and conveys his surprise at the fact that the propositional content of the higher alternative (i.e., John invited Adam, Bill and Chris) is false. The so-called "hey, wait a minute" test is first discussed in Shannon (1976) and later introduced in von Fintel (2004) as a diagnostic of speaker presupposition. Below, (15) illustrates the fact that at-issue content cannot enter into the frame "Hey, wait minute. I didn't know...", in contrast to not-at-issue content (e.g., presuppositions among others).

[3] In fact, the response in (13) can target either only the first clause in (10B) or the first clause together with the latter clause containing *at least*. The discussion here is built on the latter conversational situation. I thank one anonymous reviewer for dragging my attention to this issue.

(15) Mary's aunt is visiting today. (from Pearson 2010: (1))

a. **#Hey, wait a minute!** I didn't know Mary's aunt is visiting today.
b. **Hey, wait a minute!** I didn't know Mary has an aunt.

(16) The (not-)at-issue content in (15)

a. Assertion (at-issue): Mary's aunt is visiting today.
b. Presupposition (not-at-issue): Mary has an aunt.

Given examples like (15), the felicitous use of "Hey, wait minute" by speaker C in (14) indicates that the information content that the higher alternative is false is not-at-issue. Crucially, examples like (14) further suggest that the requirement of the concessive reading that the higher alternatives are *contextually known to be false* should be **speaker-oriented**, but not interlocutors-based (part of the common ground).

Below, (17) demonstrates that the information content that the higher alternative (i.e., John invited Adam, Bill and Chris) is false **can** be part of the common ground for the interlocutors in a given discourse. Note that like the conversation in (14), the polar question by speaker A in (17) targets the content of the prejacent. (18) # (20) illustrate three possible continuations (by speaker C) to speaker B's assertion. Notice that (18) and (19) are felicitous continuations, while (20) is not.

(17) Context: Speaker A wants to know who John invited for his party last night. All the three speakers A, B and C **know** that John didn't invite Chris. Speaker A is interested in whether the other two people (i.e., Adam and Bill) are invited.

A: Who did John invite? Did John invite Adam and Bill?
B: Yeah, **at least** he invited [Adam and Bill]$_F$.

(18) C: That's true!
(19) C: No, that's not true! John only invited Adam.
(20) C: #No, that's not true! John invited all of them.

Given the context, it is part of the common ground that the propositional content of the higher alternative (i.e., John invited Adam, Bill and Chris) is false. More specifically, the interlocutors are committed to the fact that the higher alternative is false. In (18), speaker C **assents** with speaker B with respect to the content that the prejacent (i.e., John invited Adam and Bill) is true. This indicates that the propositional content of the prejcent is at-issue. In (19), speaker C **dissents** with speaker B on the content that the prejacent is true. Notice that the lower alternative (i.e., John only invited Adam) is involved as part of speaker C's response addressing speaker A's question. This indicates that besides the prejacent, the content of the lower alternative is also at-issue. Finally, the assertion in (20) is infelicitous because it is self-contradictory: given the common ground, speaker C has committed herself to the fact that the content of the higher alternative (i.e., John invited Adam, Bill and Chris) is false, but she continues to assert that the content of the higher alternative is true.

At this point, it is worth noting that the information content regarding the ranking between the prejacent and its alternatives seems to be not-at-issue, given that it cannot be directly assented or dissented across the conversations we have seen above. For example, (18) cannot target a meaning like "John's inviting Adam, Bill and Chris is better than/ is ranked above his inviting (only) Adam and Bill". Thus, I tentatively conclude that the ranking information is not-at-issue, while the content of the prejacent is at-issue and that of its higher alternatives can be at-issue or not-at-issue.

To sum up, we have seen that for concessive *at least*, the propositional content of the prejacent is always at-issue, but the propositional content of the higher alternative can be at-issue or not-at-issue, depending on the discourse. Put differently, under the assertion with concessive *at least*, the speaker is committed to both the truth of the prejacent and the falsity of the higher alternative; however, the discourse commitment to the falsity of the higher alternative can be a result of the assertion (see (10)) or part of the common ground (see (14) and (17)). Finally, the content of the lower alternative is at-issue because it addresses the QUD (see (12) and (19)); the ranking between the prejacent and its alternatives is not-at-issue because it cannot be directly assented or dissented.

2.3 The Information Status of Epistemic *At Least*

Let us observe that under epistemic *at least*, the propositional content of the prejacent and that of its higher alternatives are both at-issue. Suppose that Adam, Bill and Chris are the three relevant individuals in the discourse. Consider (21).

(21) A: Who did John invite?

B: John invited **at least** [Adam and Bill]$_F$.

In (21), speaker A explicitly raises an issue concerning who John invited; assuming that the *wh*-question imposes an existential presupposition that John has invited someone (e.g., Dayal 2016), the domain of speaker A's question consists of a set of propositional alternatives in the structure of a semi-lattice. Next, speaker B's assertion serves as a proposal to settle the raised issue. Crucially, the issue raised by speaker A is not completely resolved under speaker B's assertion, because there are two possibilities remaining open: either John invited all the three individuals (Adam, Bill and Chris), or John (only) invited Adam and Bill. In this sense, speaker B's utterance with *at least* provides only a partial answer. (22) presents an update of the domain.[4]

[4] An anonymous reviewer points out that there is an entailment relation between the prejacent and its higher alternative in (22).There are other cases where alternatives do not involve entailment: *John at least bought [apples]$_F$* (pragmatic scale) or *John at least won a [silver]$_F$ medal* (lexical scale); however, in these cases, the fact that the speaker is committed to the possibility that the prejacent is true and the possibility that the higher alternative is true in subsequent discourse, still holds.

(22) a. The domain of speaker A's question:

$$\left\{\begin{array}{l}\lambda w.\text{ John invited}_w\text{Adam, Bill and Chris;}\\ \lambda w.\text{ John invited}_w\text{Adam and Bill; }\lambda w.\text{John invited}_w\text{Bill and Chris;}\\ \lambda w.\text{ John invited}_w\text{Adam and Chris; }\lambda w.\text{John invited}_w\text{Adam;}\\ \lambda w.\text{ John invited}_w\text{Bill; }\lambda w.\text{John invited}_w\text{Chris;}\end{array}\right\}$$

b. The domain is updated after speaker B's assertion (if accepted):
{λw.John invited$_w$ Adam, Bill and Chris; λw.John invited$_w$ Adam and Bill}

Recall that Tonhauser (2012) suggests three discourse properties of at-issue content; **(a)** at-issue content can be directly assented or dissented with; **(b)** at-issue content addresses the question-under-discussion; **(c)** at-issue content determines the relevant set of alternatives. According to properties (b) # (c) and (22), the alternatives in the domain on which epistemic *at least* is operating, namely the prejacent and its higher/ lower alternatives, all contribute to at-issue content (by the answerability test). Moreover, their discourse property of at-issueness is further confirmed by the fact that the alternatives can be directly assented or dissented, as illustrated in (23) and (24).

(23) C: That's true!
(24) C: No, that's not true! John invited only Adam.

In (23), speaker C **assents** with speaker's assertion with respect to the content that either the prejacent or its higher alternative is true. In (24), speaker C **dissents** with speaker B with respect to the content that the prejacent or its higher alternative is true. Crucially, the lower alternative (i.e., John invited (only) Adam) is involved as part of speaker C's response addressing speaker A's question. This indicates that the propositional content of the lower alternative is also at-issue. Taken together, these discourse moves indicate that the propositional content of the prejacent and those of its higher/lower alternatives are all at-issue (by the direct response test).

So far, we have seen that the information content of the alternatives on which epistemic *at least* is operating are all at-issue. Now, let us consider what the speaker's discourse commitments are under epistemic *at least*. First, the two responses in (25) and (26), where the prejacent or the higher alternative is targeted, are degraded or not justified. In (25), speaker C challenges speaker B on the content that the prejacent is true, while asserting that the content of the higher alternative is true. In (26), speaker C challenges speaker B on the content that the higher alternative is true, while asserting that the content of the prejacent is true.

(25) C: ?No, that's not true! John invited all of them/ Adam, Bill and Chris.
(26) C: ?No, that's not true! John invited (only) [Adam and Bill]$_F$.

Second, when epistemic *at least* is absent in speaker B's assertion, speaker C's challenge is akin to a contradiction. Consider (27).

(27) A: Who did John invite?

B: John invited [Adam and Bill]$_F$.
C: #No, that's not true! John invited [Adam and Bill]$_F$.

The contrast on the degradedness of the objection between (25)/(26) and (27) suggests that under the assertion with epistemic *at least*, speaker B does not *fully* commit herself to the necessary truth of the prejacent and that of the higher alternative. Instead, what speaker B is committed to seems to be the *possibility* that the prejacent is true and the *possibility* that the higher alternative is true *in subsequent discourse.* I argue that this is why epistemic *at least* has some flavor of epistemic modals. More specifically, the modal flavor does not arise at the level of lexical semantics, but at the level of pragmatics concerning the speaker's commitments in the discourse. Seen in this light, the task then is how to capture this modal flavor without hard-wiring a modal component into the meaning of SMs. Along this line of thought, the use of epistemic *at least* is typically infelicitous when the speaker **knows** that the content of the prejacent is true or the content of the higher alternative is true, as in (28) and (29).

(28) Context: Speaker B knows that John has won a silver medal in the race.
A: What medal did John win in the race?
B: #John won **at least** a [silver]$_F$ medal.

(29) Context: Speaker B knows that John has won a gold medal in the race.

A: What medal did John win in the race?
B: #John won **at least** a [silver]$_F$ medal.

In (28), the use of *at least* is infelicitous, presumably because the speaker could have been more informative (using an alternative utterance *John won a silver medal*); put differently, the possibility that John won a gold medal is unavailable in the discourse. Similarly, the use of *at least* is infelicitous in (29), presumably because the speaker could have been more informative (uttering *John won a gold medal*); again, the possibility that John won a silver medal is unavailable in the discourse. Taken together, what (25)/(26) and (28)/(29) show is that the use of epistemic *at least* is felicitous, only when the speaker is **ignorant** about whether the prejacent is true and whether its higher alternative is true (if the maxim of quantity is active in the discourse).[5] Crucially, this speaker ignorance allows

[5] There are cases where the speaker may not be ignorant when she uses epistemic *at least*. Crucially, in those cases, the maxim of quantity is *deactivated* in the discourse, e.g., the scenario of TV show repeated below; thus no ignorance inferences arise (see Mendia 2016a, b).

(i) Context: In a game, my friend has to guess the number of marbles that I have hidden. I know how many I have hidden and she knows that I have that information. I provide the clue below: I have at least five marbles.

~ > no ignorance about the number of marbles that I have.

the *possibility* that the prejacent is true and the *possibility* that the higher alternative to both project in subsequent discourse.[6]

It has been noted in previous studies that SMs are compatible with partial ignorance (Mendia 2016b, Schwarz 2016, among others). Specifically, the speaker may not be completely ignorant when she uses epistemic *at least*, as shown below.

(30) A: How many apples did John buy yesterday?

B: John bought **at least** [three]$_F$ apples.
But I know that he didn't buy {six/seven/eight...} apples.

(30) indicates that when using epistemic *at least*, the speaker may know that some of the higher alternatives are false. I believe that the compatibility of SMs with partial ignorance does not conflict with the conclusion drawn from examples like (28)/(29). In particular, when there are multiple higher alternatives (as in the numeral case above), the speaker may know that some of them, but crucially not all of them, are false in the discourse. In contrast, however, when there is only one higher alternative, as in the case of plurality scales (see (21)) or lexical scales (see (28) and (29)) above, the speaker would be ignorant about whether the higher alternative is true. Taken together, I conclude that in any case, when using epistemic *at least*, the speaker does not fully commit herself to the necessary truth of the prejacent or its higher alternative(s); instead, what she is committed to is the **projection** of the possibility that the prejacent is true and the possibility that the higher alternative/some of the higher alternatives is/are true **in subsequent discourse**.

Finally, like the case of concessive *at least*, the information content regarding the ranking between the prejacent and its alternatives seems to be not-at-issue under epistemic *at least*, as well. For example, (23) cannot target a meaning like "John's inviting Adam, Bill and Chris is better than/is ranked above his inviting (only) Adam and Bill". Therefore, I tentatively conclude that the ranking information is not-at-issue under epistemic *at least*.

To sum up, we have seen that for epistemic *at least*, the propositional content of the prejacent and its higher/lower alternatives are all at-issue, and the ranking information between the prejacent and its alternatives is not-at-issue. Finally, the use of epistemic *at least* is felicitous, only when the speaker is ignorant about whether the prejacent is

[6] The same observation applies to disjunction, where in subsequent discourse, the content of each disjunct must be (epistemically) possible to the speaker.

(i) Context: Speaker B knows that John read Hamlet yesterday.

A: What did John read yesterday?
B: #John read Hamlet or Macbeth.

(ii) Context: Speaker B knows that John read Macbeth yesterday.

A: What did John read yesterday?
B: #John read Hamlet or Macbeth.

true and whether its higher alternative is true (if the maxim of quantity is active in the discourse). Crucially, this speaker ignorance allows the *possibility* that the prejacent is true and the *possibility* that the higher alternative to both project in subsequent discourse. I consider that the projection of the two possibilities in subsequent discourse is the source of the modal flavor associated with SMs. However, I don't analyze SMs as modal expressions. Therefore, the task is then **(a)** how to capture the modal flavor without hardwiring a covert epistemic modal into the semantics of SMs; **(b)** how to capture the intuition that the speaker does not *fully* commit herself to the necessary truth of either the prejacent or its higher alternatives, and the intuition that what the speaker is committed to are the possibility that the prejacent is true and the possibility that the higher alternative is true in subsequent discourse.

2.4 Interim Summary

(31) and (32) summarize the discourse profile of concessive *at least* and epistemic *at least* for the (not-)at-issueness and the speaker's discourse commitments.

(31) **Concessive *at least***

a. The prejacent is always at-issue, while the higher alternatives can be at-issue or not-at-issue. The lower alternatives are at-issue.
b. The ranking information is not-at-issue.
c. The speaker is committed to both the falsity of the higher alternatives and the truth of the prejaent.

(32) **Epistemic *At Least***

a. The prejacent and the higher/lower alternatives are all at-issue.
b. The ranking information is not-at-issue.
c. The speaker does not *fully* commit herself to the necessary truth of either the prejacent or the higher alternatives. Instead, what the speaker is committed to is the *possibility* that the prejacent is true and the *possibility* that the higher alternative is true in subsequent discourse.

The next section presents a pragmatic analysis capturing the two different discourse profiles of *at least* under the framework of conversational scoreboard.

3 *At Least* in Conversational Scoreboard

This section proceeds as follows. Section 3.1 provides pragmatics preliminaries and introduces the discourse model (in terms of the conversational scoreboard, with the insights from Lewis 1979) presented by Farkas and Bruce (2010) and further developed in Malamud and Stephenson (2015) and Beltrama (2018). Section 3.2 #3.3 present a formal analysis of concessive *at least* and epistemic *at least*, capturing their different discourse profiles in terms of conversational scoreboard.

3.1 Pragmatic Background: Conversational Scoreboard

In this paper, I adopt the idea that discourse is structured around one or more Question-Under-Discussion (QUD; Roberts 1996/2012, Büring 2003, Beaver and Clark 2008, among others; see Sect. 4.1 on the relation between QUDs and information focus). Briefly put, a QUD is a (possibly implicit) question that amounts to a goal, at a stage of discourse: cooperative interlocutors attempt to collectively resolve the current QUD(s). In a QUD-based model, two types of core discourse moves can be made by a speaker; she can either attempt to resolve some QUD by providing an answer, or raising an issue by positing a (sub)question that could be used as a new QUD by discourse participants. For purposes of this paper and simplification, I ignore other types of discourse moves. Furthermore, I follow Biezma and Rawlins (2017a, b) and others in assuming that questioning happens against the background of a Stalnakerian common ground/context set and questions raise issues that are not settled in the context set; in contrast, answers narrow the context set and eliminate possibilities. One key idea behind the QUD-based viewpoint of discourse is the notion of *strategy*: a strategy is (roughly) a discourse path toward resolving some QUD by providing a partial/complete answer or by asking a subquestion. Finally, discourse moves must be *relevant*. In this paper, I assume Roberts (1996/2012)'s characterization of the notion of relevance, as shown below.

(33) Relevance (adapted from Roberts 1996/2012: (15))

A move M is *Relevant* to a question q iff M either introduces an (at least) partial answer to q in context c_M (M is an assertion) or is part of a strategy to answer q (M is a question).

I further assume with Roberts (1996/2012) that the notion of partial answers and complete answers are defined in terms of contextual entailment. The definition below is a modified version taken from Biezma and Rawlins (2017b: (48)).

(34) Partial answers and complete answers

a. A partial answer to a question q is a proposition which contextually entails the evaluation – either true or false – of at least one element of the alternative set characterized by q.
b. A complete answer is a proposition which contextually entails an evaluation for each element of the alternative set characterized by q.
c. p contextually entails q in a context c just in case $(p\square cs_c)$ 5 $(q\square cs_c)$, where cs_c is the Stalnakerian context set in context c.
d. p contextually entails the evaluation of q in context c iff either p contextually entails q, or p contextually entails]q.

According to the definitions above, a declarative response will be relevant if in a context, it decides (either positively or negatively) any alternative in the current QUD.

With these pragmatic preliminaries, let's turn to the discourse model presented by Farkas and Bruce (2010) (henceforth F&B).[7] A conversational state is represented in F&B's system by means of the following four main components.

(35) a. **The participants' discourse commitments** (DC_X): for each participant X, X's public discourse commitments.
b. **Table**: a stack of issues (the top issue first), where issues are represented as sets of propositions. Issues can remain on the Table only when they have been raised by previous moves and have not been resolved yet (i.e., still "under discussion").
c. **Common Ground** (CG): the set of propositions that haven already been publicly committed by all discourse participants (cf. Stalnaker 1978).
d. **Projected CG** (*CG; F&B's Projected Set): the set of *possible* CGs that give resolution(s) of the top issue on the Table in the expected next stage of the conversation (i.e., in the felicitous continuations of the conversation).

A key feature of F&B's system is that discourse moves, such as assertions and questions, are distinguished by where their associated propositions are added in the conversational scoreboard. For example, if speaker A asserts a proposition φ, then φ is added to DC_A, $\{\varphi\}$ is added to the top of the Table, and φ is added to the projected CG (i.e., CG*). If speaker B accepts the assertion (or has no objection to it), then $\{\varphi\}$ is removed from the Table and added to the CG.[8] (36) illustrates how factual assertions are computed in F&B's system.[9]

(36) (CG_i represents the input CG)

A asserts: John left.

	Before A's assertion	After A's assertion	After B accepts A's assertion
DC_A	{}	{John left}	{}
DC_B	{}	{}	{}

(*continued*)

[7] Farkas and Bruce (2010)'s model is adopted here because I believe that it is more reader-friendly and straightforward. The current analysis can be translated into an alternative discourse representation where a context is considered as a tuple consisting of various discourse components (e.g., commitment sets, the Table, CG, Projected CG, etc.) and various discourse moves are defined (e.g., push, pop, top, etc.); see Biezma and Rawlins (2017a) and Beltrama (2018) for such discourse representation.

[8] I follow the convention from F&B that when a proposition φ is added to the CG, it is also simultaneously removed from any discourse participant's commitment sets. This avoids redundancy, because the CG represents the public commitments of every discourse participant in the conversation.

[9] I follow Malamud and Stephenson (2015)'s simplification of F&B's representation of the Table. Under F&B's original representation, items placed on the Table are pairs consisting of the syntactic representation of the utterance and its denotation. Only the denotation is represented here.

(*continued*)

	Before A's assertion	After A's assertion	After B accepts A's assertion
Table	<>	<{John left}>	< >
CG	CG_i	CG_i	CG_i □ {John left}
CG*	CG_i	CG_i □ {John left}	CG_i □ {John left}

As shown above, once speaker B accepts the assertion, the new CG is obtained by intersecting the input CG and the proposition {John left} (i.e., restricting CG_i to those worlds where the proposition that John left is true; in other words, removing those worlds where John did not leave). On this view, crucially, assertions do not directly update the CG; instead, they are seen as proposals to update the input CG, available for a variety of reactions from other discourse participants.

Let's consider another case discussed in F&B: polar questions. In contrast to the assertion, a polar question creates projected CGs containing φ and those containing]φ. (37) illustrates how polar questions are computed in F&B's system.

(37) (CG_i represents the input CG)

A: Has John left?B: Yes.

	Before A's move	After A's move	After B's answer	After A accepts B's answer
DC_A	{}	{}	{}	{}
DC_B	{}	{}	{John has left}	{}
Table	< >	< {John has left, John has not left} >	< {John has left} >	< >
CG	CG_i	CG_i	CG_i	CG_i □ {John left}
CG*	CG_i	{CG_i □ {John left}, CG_i □{John has not left}}	CG_i □{John left}	CG_i □ {John left}

In F&B's system, a polar question is distinguished from an assertion in three respects. First, when a polar question is posited, what is being placed on the Table is a set consisting of the proposition φ and its complement]φ. Second, asking a polar question creates two possible projected CGs: one in which φ is added and one in which]φ is added. Third, asking a polar question does not change the speaker's discourse commitment sets.

Although F&B's system has its merit of modeling basic discourse moves such as (factual) assertions and questions, as discussed in Malamud and Stephenson (2015) and Beltrama (2018), the original system is insufficient (or not directly applicable) to model more fine-grained types of content (e.g., the interlocutors' private doxastic states; Farkas and Bruce 2010: 89) or various types of assertions (e.g., subjective assertions with a judge parameter, and assertions with tentative commitments). In the case at hand, it is not clear how to model the fact that under epistemic *at least*, the speaker does not committ herself to the necessary truth of the prejacent or the higher alternative. Notice that in F&B's system, the four components are not completely independent of each other. For example, as Malamud and Stephenson (2015: 286) has pointed out: "the commitment sets and the Table completely determine the other elements of the scoreboard: the CG consists of propositions that both (all) participants are committed to, while the projected CG consists of these joint commitments updated with all possible resolutions to the issues on the Table". In order to model cases where the speaker may not be fully committed to the propositional content that she asserts (i.e., assertions with tentative commitments), Malamud and Stephenson (2015: 288) suggest adding "Projected Commitments" into F&B's original system.

(38) **The participants' projected commitments** (DC_X*): for each participant X, X's discourse commitments in the net expected stage of the conversation (i.e., the felicitous continuations of the conversation).

Malamud and Stephenson (2015) posit sets of "projected commitments" of the speaker and the hearer(s). According to them, the modified system is able to model discourse moves giving tentative commitments (i.e., by adding propositions to the speaker's projected commitment sets) or discourse moves offering the speaker's best guess of other participants' commitments (i.e., by adding propositions to other participants' projected commitment sets).[10] An important insight behind Malamud and Stephenson (2015)'s idea of projected commitments is that adding propositions to the speaker's projected commitments leads to an inference of tentativeness. More specifically, given that the speaker is always in full control of her own commitment sets, if the speaker chooses to add propositions φ to her projected commitments but not her present commitments, then the hearer is entitled to infer that the speaker has some reason to delay making her present commitments. With no other obvious pragmatic reasons, typically, the hearer can infer that *the speaker thinks* φ*, but she is uncertain about* φ. This licenses an inference of tentativeness.

I follow Malamud and Stephenson (2015) in adding "projected commitments" into the conversational scoreboard. For purposes of this paper, the hearer's projected commitments will not concern us. Instead, what matter to us will be the speaker's projected commitments. As we will see shortly, I suggest that the tentativeness is the pragmatic source of the modal flavor associated with epistemic *at least*. To anticipate, my proposal in a nutshell is that although concessive *at least* and epistemic *at least* share one single semantic representation, they have different discourse profiles and the discrepancy in their discourse profiles crucially comes from where propositions are added in the conversational scoreboard. More specifically, assertions with concessive *at least* add propositions to the speaker's present commitments, while assertions with epistemic *at least* add propositions to the speaker's projected commitments.

[10] There is an asymmetry between regular present commitments (DC_X) and projected commitments (DC_X*): a discourse move may add propositions to either the speaker' or the hearer's projected commitment sets, but it can only add propositions to the speaker's (but crucially not the hearer's) regular present commitments.

The next section presents a formal analysis of concessive *at least* capturing its discourse profile and illustrates how assertions with concessive *at least* are computed in terms of conversational scoreboard.

3.2 Concessive *At Least*

Let's first consider the case where the content of the prejacent and that of the higher alternative are both at-issue. The relevant example is repeated below.

(39) A: Who did John invite? Did John invite Adam, Bill and Chris?
B: No, John didn't invite Adam, Bill and Chris.

But, **at least** he invited [Adam and Bill]$_F$.

To address the QUD (the issue raised by speaker A whether John invited Adam, Bill and Chris), speaker B first makes an assertion that John didn't invite Adam, Bill and Chris, and then makes another assertion with concessive *at least* that John invited Adam and Bill. The former assertion amounts to a proposal to update the CG with the content that the higher alternative is false, and the latter assertion a proposal to update the CG with the content that the prejacent is true. With this in mind, the conversational scoreboard in (40) illustrates how assertions with concessive *at least* are computed when the content of the prejacent and the higher alternative is at-issue.

(40) (CG_i represents the input CG;
$q\ B_c\ p$ represents that q is ranked higher than p in context c)

A asserts $]q$ and then asserts p with concessive *at least*

	Before the move	After the move
DC_A	{}	$\{\{]q\}, \{p\}\}$
DC_A*	{}	{}
DC_B	{}	{}
Table	< >	$<\{]q\}, \{p\}>$
CG	CG_i	$CG_i \Box \{q\ B_c\ p\}$
CG*	CG_i	$(CG_i \Box \{q\ B_c\ p\}) \Box \{]q\} \Box \{p\}$

In (40), the speaker makes two proposals to update the CG: $]q$ and p. Therefore, two propositions (i.e., $]q$ and p) are added to the speaker's present commitments and two items (i.e., $\{]q\}$ and $\{p\}$) are placed on the Table. Assume that not-at-issue content directly updates the input CG without being subject to the acceptance of other discourse participants (e.g., AnderBois et al. 2010, 2015 on clause-medial appositives and Murray 2014 on evidentials), the information content of the ranking (i.e., $q\ B_c\ p$) is directly added into the CG (i.e., $CG_i \Box \{q\ B_c\ p\}$).[11] Finally, the projected CG is obtained by intersecting

[11] One anonymous reviewer points out that not all not-at-issue content can update CG (e.g., presuppositions); they need to be new information. I leave open the possibility that ranking information is actually a presupposition in the discourse.

the input CG (with the added not-at-issue content) and the two propositions (i.e., $(CG_i \Box\{q\ B_c\ p\}) \Box \{]q\} \Box \{p\})$. Given the conversational scoreboard in (40), the concessive meaning comes from the fact that q is higher ranked than p in the context c (with some additional pragmatic attachment of preference), but q is false and p is true. Moreover, it captures our intuition that under concessive *at least*, the speaker is committed to both the falsity of the higher alternative and the truth of the prejacent, because both $\{]q\}$ and $\{p\}$ are in the speaker's present commitments.

Next, let's turn to the case where the content of the prejacent is at-issue, but that of the higher alternative is not-at-issue. The relevant example is repeated below.

(41) Context: Speaker A wants to know who John invited for his party last night. All the three speakers A, B and C **know** that John didn't invite Chris. Speaker A is interested in whether the other two people (i.e., Adam and Bill) are invited.

(42) A: Who did John invite? Did John invite Adam and Bill?
B: Yeah, **at least** he invited [Adam and Bill]$_F$.

In contrast to the previous example, to address the QUD (the issue raised by speaker A whether John invited Adam and Bill), speaker B makes only one assertion with concessive *at least* that John invited Adam and Bill. The conversational scoreboard in (43) illustrates how assertions with concessive *at least* are computed when the content of the prejacent is at-issue but that of the higher alternative is not-at-issue.

(43) (CG_i represents the input CG and $\{]q\}$ is part of CG_i;

$q\ B_c\ p$ represents that q is ranked higher than p in context c)
A asserts p with concessive *at least*

	Before the move	After the move
DC_A	$\{\}$	$\{p\}$
DC_A*	$\{\}$	$\{\}$
DC_B	$\{\}$	$\{\}$
Table	$< >$	$< \{p\} >$
CG	CG_i	$CG_i \Box \{q\ B_c\ p\}$
CG*	CG_i	$(CG_i \Box \{q\ B_c\ p\}) \Box \{p\}$

In (43), the speaker makes a proposal to update the CG: p. Therefore, the proposition p is added to the speaker's present commitments and $\{p\})$ is placed on the Table. The information content of the ranking (i.e., $q\ B_c\ p$) is not-at-issue and is thus directly added into the CG (i.e., $CG_i \Box \{q\ B_c\ p\}$). The projected CG is obtained by intersecting the input CG (with the added not-at-issue content) and the proposition p (i.e., $(CG_i \Box \{q\ B_c\ p\}) \Box \{p\})$. Notice that $\{]q\}$ is already part of CG_i. This is why the content of the higher alternative q is not-at-issue. Finally, given the conversational scoreboard in (43), the concessive meaning arises from the fact that q is higher ranked than p in the context

c (with some additional pragmatic attachment of preference), but q is false and p is true. (43) also captures our intuition that under concessive *at least*, the speaker is committed to both the falsity of the higher alternative and the truth of the prejacent, because $\{]q\}$ is part of CG and $\{p\}$ is in the speaker's present commitments.

3.3 Epistemic *At Least*

Recall that the use of epistemic *at least* is felicitous only when the speaker is ignorant about whether the prejacent is true and whether its higher alternative is true (if the maxim of quantity is active in the discourse). Crucially, this speaker ignorance allows the *possibility* that the prejacent is true and the *possibility* that the higher alternative to both project in subsequent discourse. This projection behavior of the two possibilities in subsequent discourse is considered as the source of the modal flavor associated with SMs. Therefore, the main task of this section is to illustrate **(a)** how to capture the modal flavor of epistemic *at least* (more generally, SMs) at the level of discourse; **(b)** how to capture the intuition that the speaker does not *fully* commit herself to the necessary truth of either the prejacent or the higher alternatives, and what the speaker is committed to are the possibility that the prejacent is true and the possibility that the higher alternative is true in subsequent discourse. A canonical example of epistemic *at least* is repeated below.

(44) A: Who did John invite? B: John invited **at least** [Adam and Bill]$_F$.

To address the QUD (the issue raised by speaker A concerning the individuals invited by John), speaker B makes an assertion with epistemic *at least* that John invited Adam and Bill. Crucially, the speaker's assertion provides only a partial answer, because two possibilities remain open in the discourse: (a) John invited Adam, Bill and Chris (i.e., the higher alternative); (b) John invited (only) Adam and Bill (i.e., the prejacent). With the above in mind, (45) illustrates how assertions with epistemic *at least* are computed under the conversational scoreboard.

(45) (CG_i represents the input CG;

$q\ B_c\ p$ represents that q is ranked higher than p in context c)
Aasserts p with epistemic *at least*

	Before the move	After the move
DC_A	$\{\}$	$\{\}$
DC_A*	$\{\}$	$\{\{q\}, \{p\}\}$
DC_B	$\{\}$	$\{\}$
Table	$< >$	$< \{q\}, \{p\} >$
CG	CG_i	$CG_i \square \{q\ B_c\ p\}$
CG*	CG_i	$\{(CG_i \square \{q\ B_c\ p\}) \square \{q\}\}, (CG_i \square \{q\ B_c\ p\}) \square \{p\}\}$

In (45), the speaker makes two tentative proposals to update the CG: q and p. Therefore, two items (i.e., $\{q\}$ and $\{p\}$) are placed on the Table. However, two propositions (i.e., q and p) are added to the speaker's projected commitments, but crucially not to the speaker's present commitments. Assume that not-at-issue content directly updates the input CG, the information content of the ranking (i.e., $q\ \mathrm{B}_c\ p$) is directly added into the CG (i.e., $\mathrm{CG}_i \Box q\ \mathrm{B}_c\ p\}$). Finally, the assertion with epistemic *at least* creates two possible projected CGs: one in which $\{q\}$ is added is obtained by intersecting the input CG (with the added not-at-issue content) and the proposition q (i.e., $(\mathrm{CG}_i \Box p\{q\ \mathrm{B}_c\}) \Box\{q\}$); one in which $\{p\}$ is added is obtained by intersecting the input CG (with the added not-at-issue content) and the proposition p (i.e., $(\mathrm{CG}_i \Box \{q\ \mathrm{B}_c\ p\}) \Box \{p\}$). Given the conversational scoreboard in (45), the modal flavor of epistemic *at least* comes from the fact that both propositions p and q (the prejacnet and its higher alternative) are projected in subsequent discourse, more specifically, in two different projected CGs. Furthermore, it also captures the intuition that under epistemic *at least*, the speaker does not *fully* commit herself to the necessary truth of either the prejacent or the higher alternatives (i.e., two items $\{q\}$ and $\{p\}$ are placed in DC_A*); instead, what the speaker is committed to are the possibility that the prejacent is true and the possibility that the higher alternative is true in subsequent discourse (i.e., $\{(\mathrm{CG}_i \Box \{q \Box_c p\}) \Box \{q\}\}$, $(\mathrm{CG}_i \Box \{q \Box_c p\}) \Box\{p\}\}$).

It is worth emphasizing that according to (45), assertions with epistemic *at least* are both **informative** and **inquisitive**, namely, they have mixed properties of both assertions and questions. On the one hand, they are informative because they are like (factual) assertions in **(a)** updating the speaker's commitment sets and **(b)** serving as a proposal to update the CG (subject to other discourse participants' acceptance). On the other hand, they are inquisitive because they are like (polar) questions in creating multiple possible projected CGs.

To appreciate the mixed properties of those assertions with epistemic *at least*, let's briefly recall how assertions and polar questions are modeled in F&B's system. For (factual) assertions, if speaker A asserts a proposition φ, then φ is added to DC_A, $\{\varphi\}$ is added to the top of the Table, and φ is added to the projected CG (i.e., CG*). If speaker B accepts the assertion (or has no objection to it), then $\{\varphi\}$ is removed from the Table and added to the CG. In contrast, a polar question is distinguished from an assertion in three respects. First, when a polar question is posited, what is being placed on the Table is a set consisting of the proposition φ and its complement $]\varphi$. Second, asking a polar question creates two possible projected CGs: one in which φ is added and one in which $]\varphi$ is added. Third, asking a polar question does not change the speaker's discourse commitment sets.

Finally, I would like to highlight several important features of the current analysis: **(a)** the modal flavor associated with epistemic *at least* is *not* at the level of lexical semantics (i.e., not via a covert epistemic modal), but lies in the level of discourse; **(b)** assertions with epistemic *at least* add associated propositions to the speaker's projected commitments, crucially *not* to the speaker's present commitments. This delay in making a commitment (resulting in an inference of tentativeness/ uncertainty) constitutes one pragmatic source of the modal flavor; **(c)** under an assertion with epistemic *at least*, multiple possible projected CGs are generated in the discourse, which constitutes another pragmatic source of the modal flavor; (**d**) assertions with epistemic *at least* are both informative and inquisitive.

4 Chinese *Zhishao* 'At Least'

Two key observations about EPI and CON (among others) discussed in Sect. 2 are: (**a**) Under CON, the speaker is committed to both the truth of the prejacent and the falsity of the higher alternatives; crucially, the latter information can be either *at-issue* (i.e., asserted by the speaker) or *not-at-issue* (e.g., part of the common ground); (**b**) Under EPI, the speaker is only committed to the *possibility* that the prejacent is true and the *possibility* that some higher alternative is true in *subsequent discourse*. Below, I show that these discourse properties also hold for Chinese sentences with *zhishao* under the two readings, which implicates that the proposed pragmatic analysis can be extended to expressions in other languages such as *zhishao*.

First, (46) illustrates the fact that Chinese *zhishao* demonstrates the familiar EPI-CON ambiguity: Under CON, (46) conveys that *Zhangsan* has written three novels, though it is not the optimal but acceptable. Under EPI, (46) conveys that the speaker is not sure exactly how many novels that *Zhangsan* has written.

(46) Zhangsan **zhishao** xie-le [san]$_F$-ben-xiaoshuo. **EPI, CON**
Zhangsan at least write-ASP three-CL-novel
'Zhangsan at least wrote three novels.'

Second, (47) shows that the falsity of the higher alternative can be simply at-issue (i.e., inviting all pop stars we want), explicitly asserted by the speaker; by contrast, the information content in question can be not at-issue (i.e., John did not invite everyone/Chris), part of the common ground, as in (48).

(47) Suiran Zhangsan meiyou yaoqing suoyou mingxin,
Although Zhangsan did.not invite all pop-stars
Zhishao ta yaoqing-le [Yadang he Bier]$_F$.
at least he invite-ASP Adam and Bill
'Although Zhangsan did not invite all pop stars (we want), at least he invited Adam and Bill'

(48) Context: Speaker A wants to know who John invited for his party last night Both speakers A and B **know** that John didn't invite Chris. Speaker A is interested in whether the other two people (i.e., Adam and Bill) are invited.

A: In the end, who did John invite? Did John invite Adam and Bill?
B: Shi-a... , **zhishao** Yuehan yaoqing-le [Yadang he Bier]$_F$.
Be-SFP at least John invite-ASP Adam and Bill
Xietianxiedi!
Thank.God!
'Yeah....at least John invited Adam Bill. Thank God!'

Finally, (49) and (50) illustrate that in the conversation where *zhishao* (under EPI) is used to signal the possibilities that the prejacent or the higher alternative is true in subsequent discourse; it seems *infelicitous* (though not a contradiction) to challenge the necessary truth of the prejacent and that of the higher alternative. Note that **without** ***zhishao*** **in (49)**, the challenge in (50a) would simply sound like a *contradiction*.

(49) A: What medal did Zhangsan win in the race?
B: Zhangsan **zhishao** na-le [yin]$_F$-pai.
Zhangsan at least take-ASP silver-medal
'Zhangsan at least got a silver medal.'

(50) a. #C: Ni cuo-le! Zhangsan (zhi) na-le [yin]$_F$-pai!
You wrong-ASP Zhangsan only take-ASP silver-medal
'You are wrong! Zhangsan (only) got a silver medal'
b. #C: Ni cuo-le! Zhangsan na-le [jin]$_F$-pai!
You wrong-ASP Zhangsan take-ASP gold-medal
'You are wrong! Zhangsan got a gold medal!'

The above discussion indicates that the pragmatic analysis based on conversational scoreboard presented in this paper can be extended to Chinese *zhishao* as well, to capture its different discourse profiles under the two readings. For reasons of space, I leave the detailed formal implementation for another occasion.

5 Conclusions

This paper first considered how an assertion with *at least* under two readings (EPI and CON) fares with respect to diagnostics of (not-)at-issue content and updates the discourse, and then presented a pragmatic analysis capturing their different discourse profiles under the system of conversational scoreboard in Farkas and Bruce (2010) and subsequently developed by others (e.g., Malamud and Stephenson 2015). In particular, it was shown that the speaker is committed to both the truth of the prejacent and the falsity of the higher alternatives under CON. By contrast, under EPI, the speaker is only committed to the *possibility* that the prejacent is true and the *possibility* that some higher alternative is true in *subsequent discourse*. This modal flavor of *at least* arises at the level of discourse, rather than lexical semantics (e.g., via a covert epistemic modal). Finally, it was suggested that the pragmatic analysis can be extended to expressions in other languages such as Chinese *zhishao* 'at least', which also demonstrates the familiar EPI-CON ambiguity.

References

AnderBois, S., Brasoveanu, A., Henderson, R.: Crossing the appositive / at-issue meaning boundary. In: Li, N., Lutz, D. (eds.) Proceedings of Semantics and Linguistic Theory 20, pp. 328–346 (2010)

AnderBois, S., Brasoveanu, A., Henderson, R.: At-issue proposal and appositive impositions in discourse. J. Semant. **32**, 93138 (2015)

Beltrama, A.: Totally between subjectivity and discourse. Exploring the pragmatic side of intensification. J. Semantics **35,** 219–261 (2018)

Biezma, M.: Only one *at least*. In: Proceedings of Penn Linguistics Colloquium, (ed.) Kobey Shwayder, vol. 36, pp. 12–19 (2013)

Biezma, M., Rawlins, K.: Rhetorical questions: Severing questioning from asking. In: Burgdorf, D., Collard, J. (eds.) Proceedings of Semantics and Linguistic Theory 27, LSA and CLC Publications (2017a)

Biezma, M., Rawlins, K.: Or what?. *Semantics and Pragmatics* (2017b)

Büring, D.: On D-trees, beans, and B-accents. Linguist. Philos. **26**, 511–545 (2003)

Coppock, E., Brochhagen, T.: Raising and resolving issues with scalar modifiers. Semantics and Pragmatics **6**, 1–57 (2013)

Dayal, V.: Questions. Oxford University Press, Oxford (2016)

Farkas, D., Bruce, K.B.: On reacting to assertations and polar questions. J. Semant. **27**, 81–118 (2010)

von Fintel: Would You Believe It? The King of France is Back! (Presuppositions and Truth-Value Intuitions). In: Descriptions and Beyond, 315–341. Oxford University Press, Oxford (2004)

Kay, P.: At least. In: Lehrer, A., Kittay, E.F. (eds.) Frames, Fields, and Contrasts: New Essays in Semantic and Lexical Organization, pp. 309–331. Lawrence Erlbaum Associates, Hillsdayle, NJ (1992)

Koev, T.: On the information status of appositive relative clauses. In: Proceedings of Amsterdam Colloquium, vol. 18, pp. 401–410, Berlin: Springer (2012)

Koev, T.: Apposition and the Structure of Discourse. Doctoral Dissertation, Rutgers University, New Brunswick (2013)

Lewis, D.: Scorekeeping in a language game [Reprinted 2002]. In: Portner, P., Partee, B.H. (eds.) Formal Semantics: The Essential Readings, pp. 162–177. Blackwell, Oxford (1979)

Malamud, S., Stephenson, T.: Three ways to avoid commitments: declarative force modifiers in the conversational scoreboard. J. Semant. **32**, 275–331 (2015)

McCready, E.: What man does. Linguist. Philos. **31**(6), 671–724 (2009). https://doi.org/10.1007/s10988-009-9052-7

McCready, E.: Varieties of conventional implicature. Semant. Pragmatics **3**, 1–57 (2010)

Mendia, J.A.: Focusing on Scales. In: Proceedings of North East Linguistic Society (NELS) 46, Concordia University, Montreal (2016a)

Mendia, J.A.: Reasoning with partial orders: restrictions on Ignorance Inferences of Superlative Modifiers. In: Proceedings of Semantics and Linguistic Theory 26, University of Texas Austin (2016b)

Murray, Sarah E.: *Evidentiality and the structure of speech acts*. Doctoral Dissertation, Rutgers University, New Brunswick (2010)

Murray, S.E.: Varieties of update. Semantics and Pragmatics **7,** 1–53 (2014)

Nakanishi, Kimiko, and Hotze Rullmann. 2009. Epistemic and Concessive Interpretation of at least. In CLA 2009

Pearson, H.: A modification of the"Hey, Wait a Minute" test. Snippets **22**, 7–8 (2010)

Potts, C.: The Logic of Conventional Implicatures. Oxford University Press, Oxford (2005)

Potts, C.: The expressive dimension. Theoretical Linguistics 33(2), 165–198 (2007). 271–297. Blackwell Publishers, Oxford

Roberts, C.: Information structure in discourse: Towards an integrated formal theory of pragmatics. Semantics Pragmatics **5**, 1–69 (1996/2012)

Schwarz, B.: Consistency preservation in Quantity implicature: the case of *at least*. Semant. Pragmatics **9**, 1–47 (2016)

Shannon, B.: On the two kinds of presuppositions in natural language. Found. Lang. **14**, 247–249 (1976)

Simons, M., Tonhauser, J., Beaver, D., Roberts, C.: What projects and why. In: Li, N., Lutz, D. (eds.) Proceedings of Semantics and Linguistic Theory 20, pp. 309–327 (2010)

Stalnaker, R.: Assertion. In: Peter Cole (ed.) Syntax and Semantics, New York: Academic Press, volume 9: Pragmatics, 315–332 (1978)

Syrett, K., Koev, T.: Experimental evidence for the truth conditional contribution and shifting information status of appositives. J. Semant. **32**, 525–577 (2015)

Tonhauser, J.: Diagnosing (not-)at-issue content. In: Proceedings of Semantics of Under-Represented Languages of the Americas 6, E. Bogal-Allbritten (ed.), 239–254 (2012)

Tonhauser, J., Beaver, D., Roberts, C., Simons, M.: Toward a taxonomy of projective content. Language **89**, 66–109 (2013)

Disjunction in a Predictive Theory of Anaphora

Patrick D. Elliott(✉)

Heinrich Heine University Düsseldorf, Düsseldorf, Germany
patrick.d.elliott@gmail.com

Abstract. In this paper I develop a dynamic semantics for a first-order fragment, which incorporates insights from work on anaphora in logic, and the trivalent approach to presupposition projection. The resulting system—EDS—has interesting features which set it apart, both conceptually and empirically, from earlier iterations of dynamic semantics. Conceptually, the meanings of the logical connectives are derived by systematically generalizing the Strong Kleene connectives into a dynamic setting—the system is thereby *predictive*, drawing a tight connection between the logic of presupposition projection and patterns of anaphoric accessibility. On the empirical side, EDS diverges sharply from earlier proposals. In this paper, I focus mainly on disjunction, arguing that EDS provides a simple and elegant account of the dynamics of disjunction, including traditionally problematic cases such as Partee disjunctions and program disjunctions.

Keywords: Disjunction · Presupposition · Anaphora

1 Introduction

Dynamic theories of natural language semantics traffic in anaphoric information. Pretty much everyone agrees that indefinites and pronouns are special—indefinites *introduce* anaphoric information—Karttunen's *discourse referents* [1]—and pronouns *retrieve* anaphoric information. A pertinent question arises: to what extent to one needs to make reference to anaphoric information in the semantics of other expressions, such as logical vocabulary (*and*, *or*, etc.)? Many proposals submit that logical expressions may arbitrarily encode a complex set of instructions for regulating the flow of anaphoric information [2,3].

In this paper, I'll develop a different kind of dynamic semantics, taking anaphoric dependencies in disjunctive sentences as a case study. I'll maintain the idea that indefinites and pronouns are special, but I'll explore the possibility that we can make use of independently motivated machinery for explaining *presupposition projection*—concretely, the Strong Kleene logic of indeterminacy—in order to help us understand *why* different logical expressions regulate anaphoric information in just the way that they do.

In Sect. 2 I provide a brief précis of Dynamic Predicate Logic (DPL) [3]. This will serve two purposes:

D. Deng et al. (Eds.): TLLM 2022, LNCS 13524, pp. 76–98, 2023.
https://doi.org/10.1007/978-3-031-25894-7_4

1. The Strong Kleene dynamic logic which I develop in this paper will make use of notions first made precise in DPL.
2. DPL will serve as a good representative of (a certain family of) dynamic theories of natural language semantics.

In Sect. 3 I'll discuss empirical challenges for DPL, focusing on the internal and external dynamics of disjunction. This naturally leads into Sect. 4, where I develop a new logic of anaphora: EDS. This logic is based on the idea that it's possible to embed the logic of presupposition projection into a dynamic setting by computing three DPL-style meanings in tandem, corresponding to the three truth values of trivalent logic. In Sect. 5, I embed EDS in a concrete discourse pragmatics, by adopting Heim's notion of an information state [2], together with a concrete bridge principle. This will be essential in order to understand how the permissiveness of EDS might be reigned in. Finally, in Sect. 6, I briefly survey some recent, related approaches to anaphora before concluding.

2 Dynamic Predicate Logic

DPL [3] provides a dynamic interpretation for a simple first-order calculus. The interpretation of a sentence is a *relation between assignments*—I will assume some familiarity with DPL in this paper, so the presentation will remain rather terse. Before sketching out the details, note that I depart in a couple of notable ways from the presentation of Groenendijk and Stokhof [3] (henceforth G&S). Firstly, following, e.g., van den Berg's presentation, *discourse referent introduction* is cashed out as random assignment. Furthermore, DPL interpretations are stated relative to a world of evaluation.[1]

(1) **Dynamic Predicate Logic:**
  a. if ϕ is atomic, then $[\![\phi]\!]^w := \{\, (g,h) \mid g = h \wedge [\phi]^{w,g} \text{ is } \mathbf{true} \,\}$
  b. $[\![\varepsilon_v]\!]^w := \{\, (g,h) \mid g[v]h \,\}$
  c. $[\![\phi \wedge \psi]\!]^w := [\![\phi]\!]^w \circ [\![\psi]\!]^w$
  d. $[\![\neg\phi]\!]^w := \{\, (g,h) \mid g = h \wedge \{\, i \mid (g,i) \in [\![\phi]\!]^w \,\} = \emptyset \,\}$

Atomic sentences (1a) are *tests*, i.e., they do not introduce anaphoric information, but merely assess (classical) truth with respect to an assignment g. Random assignment (1b) is responsible for introducing anaphoric information: ε_v is a privileged tautology, which in a dynamic setting means that for every assignment g, there is some assignment h, s.t., $(g,h) \in [\![\varepsilon_v]\!]^w$. Concretely, ε_v indeterministically assigns a value to v—$g[v]h$ holds just in case g and h differ at most in the value they assign to v. Random asignment is used to introduce discourse referents, which are threaded from left-to-right via dynamic conjunction (1c), which is just relational composition.[2] For example, the sentence "there

[1] This will later prove useful when embedding EDS in a concrete discourse pragmatics.
[2] The definition of relational composition (which is totally standard) is given below:

$$R \circ S := \{(g,i) \mid \exists h[(g,h) \in R \wedge (h,i) \in S]\}$$

This operation plays a central role in both DPL and EDS.

is a bathroom" is translated into DPL as $\varepsilon_v \wedge B(v)$. The open sentence $B(v)$ is interpreted relative to the discourse referent introduced by random assignment, thereby narrowing down assignments just to those that map v to a bathroom in w. This is illustrated below in (2).

(2) $[\![\varepsilon_v \wedge B(x)]\!]^w$
  a. $= [\![\varepsilon_v]\!]^w \circ [\![B(x)]\!]^w$
  b. $= \{ (g, h) \mid g[v]h \} \circ \{ (g, h) \mid g = h \wedge g(v) \in I_w(B) \}$
  c. $= \{ (g, h) \mid g[v]h \wedge h(v) \in I_w(B) \}$

More generally conjunction in DPL is associative, thanks to the associativity of relational composition:

(3) **Associativity of dynamic conjunction (DPL):**
$\phi \wedge (\psi \wedge \sigma) \iff (\phi \wedge \psi) \wedge \sigma$

Associativity, together with random assignment, underlies *Egli's theorem*, which ecapsulates the DPL account of discourse anaphora: "there is a^v bathroom, and it$_v$'s upstairs" is semantically equivalent to "there is a^v bathroom upstairs" in DPL.

(4) **Egli's theorem (DPL):**
$\varepsilon_v \wedge (\phi \wedge \psi) \iff (\varepsilon_v \wedge \phi) \wedge \psi$

Dynamic negation (1d) will have an important role to play in the following discussion, so it's worth dwelling on. The definition of negation in DPL is tailored to ensure that negative sentences are *anaphorically inert*. Concretely, negative sentences are *tests*, which succeed if the scope ϕ is false with respect to an assignment. This means that placing, e.g., (2) in the scope of negation results in a sentence that is (a) anaphorically inert, and (b) has the truth-conditions of a negative existential statement.

(5) $[\![\neg(\varepsilon_v \wedge B(v))]\!]^w$
  a. $= \{ (g, h) \mid g = h \wedge \{ i \mid (g, i) \in [\![\varepsilon_v \wedge B(v)]\!]^w \} = \emptyset \}$
  b. $= \{ (g, h) \mid g = h \wedge I_w(B) = \emptyset \}$

Existential quantification is defined syncategorematically in terms of random assignment and dynamic conjunction (6a). Disjunction (6b) and implication (6c) are defined syncategorematically in terms of classically equivalent sentences. A hallmark of DPL is that it matters exactly *which* classical equivalences one uses to define these connectives—the choise is *crucial* for constraining the flow of anaphoric information in complex sentences.

(6) **DPL definitions**
  a. $\exists_v \phi := \varepsilon_v \wedge \phi$
  b. $\phi \vee \psi := \neg(\neg\phi \wedge \neg\psi)$
  c. $\phi \rightarrow \psi := \neg(\phi \wedge \neg\psi)$

Since disjunction is the primary focus of this paper, I'll focus on (6b). Since a disjunctive sentence is a negative sentence it is anaphorically inert—a discourse referent introduced in a disjunction cannot be retrieved by a subsequent open sentence. To use DPL terminology, disjunction is *externally static* (by way of contrast, conjunction is externally *dynamic*). This is taken to be desirable, since as noted by G&S, anaphora from out of a disjunctive sentence is seemingly impossible in natural language:[3]

(7) Either this house is derelict, or there's a^v bathroom. #It_v's upstairs.

Furthermore, since each disjunct is itself a negative sentence, each disjunct is anaphorically inert, and therefore anaphora between disjuncts is impossible. In DPL terminology, disjunction is *internally static* (again, conjunction in contrast is internally dynamic). Similarly, G&S suggest that this is desirable on the basis of natural language—the following example is from [4, p. 245].

(8) #Either Jones owns a^v bicycle, or it_v's broken.

Implication will not be the main focus of this paper, but briefly—DPL implication is tailored to derive *universal* readings for the famous case of Donkey Sentences. I.e., (9) is taken to be equivalent to *every bathroom is upstairs.* More generally, *Egli's corollory* holds in DPL.[4] With regards to anaphora, implication is internally dynamic but externally static.

(9) If there's a^v bathroom, it_v's upstairs.

(10) **Egli's corollary**
$\exists_x \phi \rightarrow \psi \iff \forall_x(\phi \rightarrow \psi)$

DPL provides a simple and elegant logic of anaphoric information, but the way in which disjunction and implication are defined is rather *ad hoc*—as noted previously, it really matters which classical equivalences one uses to define disjunction and implication, and which operations one takes to be basic. For example, it wouldn't do to take dynamic disjunction to be a basic operation, and thereby define conjunction as $\neg(\neg\phi \vee \neg\psi)$ via de Morgan's equivalence. This would predict that anaphora should be impossible between conjuncts, despite the fact that the truth-conditional import of such a conjunction would be classical. There is therefore no obvious way of *determining* the DPL semantics of a logical operator given its truth-conditional import.

Furthermore, there are a multitude of possible connectives that are definable in DPL which would manipulate anaphoric information in a way which wouldn't correspond to their language counterparts. I take that it would be desirable to have a general recipe for determining how logical connectives manipulate anaphoric information, on the basis of their truth-conditional contribution.

---

[3] G&S importantly assume that a multi-sentence discourse is translated into DPL as a conjunctive sentence.

[4] N.b. the universal quantifier is defined as the dual of the existential.

This is related to the explanatory problem for dynamic semantics, discussed most frequently with respect to Heim's satisfaction theory presupposition projection [5] (see, e.g., [6–9]). The conceptual problem is equally acute in the case of anaphora, and unlike the case of presupposition projection, fewer alternatives have been explored.[5]

3 Empirical Challenges

3.1 Double Negation

It has been continuously pointed out, including by G&S themselves and in much subsequent work, that the empirical predictions made by DPL often do not match up with our intuitions about natural language. The most straightforward challenge is that, in DPL, Double Negation Elimination (DNE) isn't valid. This is because, since any negative sentence is a test, a doubly-negated sentence is always a test. However, as pointed out by Krahmer and Muskens [15] and Gotham [16] among others, doubly-negated sentences often license anaphora. The following example is from [15].

(11) It's not true that John didn't bring an^v umbrella.
It$_v$ was purple and stood in the hallway.

One desideratum of the account developed in Sect. 4 is to have a dynamic logic in which DNE *is* valid. There are some potential objections which are worth immediately addressing. Gotham [16] suggests that the facts are more nuanced, in that doubly-negated sentences may carry inferences that their positive counterparts lack. His example is given in (12). His point is that this discourse sounds odd because it implies that John owns a single shirt; the conclusion is that $\neg\neg\exists_v \phi$ carries a uniqueness inference that $\exists_v \phi$ lacks.

(12) It's not true that John doesn't own a^v shirt. ?It$_v$'s in the wardrobe.

I however agree with Mandelkern (cited as p.c. in [16]) that it's possible to demonstrate that doubly-negated sentences do not systematically entail uniqueness. This can be shown by using a generalization of Heim's famous sage plant example.

(13) It's not the case that Sue didn't buy a^v sage plant.
In fact, she bought eight others along with it$_v$!

[5] As emphasized by Mandelkern and Rothschild [10] the kind of situation-based e-type approach to anaphora developed in [11] and refined in [12,13] does *not* (in its current state, at least) consitute a viable alternative. E-type theories have not addressed in detail how to capture notions of anaphoric accessibility in complex sentences, beyond donkey sentences, and as shown in [14], were they to do so, they would require entries for the logical connectives which manipulate minimal situations in an apparently arbitrary fashion.

Moreover, the minimal positive counterpart of (12) to my ear also strongly implies that John only owns a single shirt. It's certainly an interesting question to ask how exactly such uniqueness inferences arise, but for the purposes of this paper, I'll be setting them to one side.

(14) John does own a^v shirt. ?It$_v$'s in the wardrobe.

Besides, developing a dynamic logic in which DNE is valid will have positive ramifications elsewhere, for example in the treatment of disjunction. I take it that developing a dynamic logic in which DNE is valid is a reasonable starting point; doubly-negated sentences undoubtedly differ from their positive counterparts in certain respects, but this is somewhat unsurprising, especially from a Gricean perspective.[6] Having outlined the problems associated with negation in DPL, I now turn to the main focus of this paper: disjunction.

3.2 Partee Disjunctions

As noted, DPL disjunction is internally static. A famous example originally due to Barbara Partee (henceforth: *Partee disjunctions*) suggests that this isn't quite right for natural language. (15) is in a sense doubly surprising in the context of DPL, since as well as seemingly involving an anaphoric dependency between disjuncts, it also seemingly involves anaphoric information introduced by a negative sentence (the first disjunct).

(15) Either there's no^v bathroom, or it$_v$'s upstairs.

At this stage, it's worth establishing some desiderata for the eventual treatment of Partee disjunctions, since there is some disagreement in the literature on their truth-conditions. For example, [15] suggests that (15) has universal truth conditions, by analogy with the DPL treatment of donkey sentences. Their analysis predicts that (15) implies that *every bathroom is upstairs*. Related to the discussion of double negation, Gotham claims that (15) carries a conditional uniqueness inference, i.e., *if there is a bathroom, then there is exactly one*. Even if universal/uniqueness readings exist, I argue here that both are at least sometimes too strong. Much like donkey sentences,[7] Partee disjunctions can have existential readings. (16) is true just in case (a) Gabe has no credit card, (b) Gabe has at least one credit card and paid with one of his credit cards. Crucially for the present point, (16) is true if Gabe has a credit card he paid with, and one that he didn't.

[6] If ϕ and $\neg\neg\phi$ are equivalent, then choosing to use a sentence of the form $\neg\neg\phi$ is naturally expected to trigger a *Manner* implicature. I leave the interesting question of the pragmatics of doubly-negated sentences to future work.

[7] (16) is in fact modelled after the following well-known example used to motivate existential readings of donkey sentences (attributed by [17, p. 63] to Robin Cooper).

(1) Yesterday, every person who had a credit card paid his Bill with it.

(16) Either Gabe doesn't have a credit card, or he paid with it.

(16) is already incompatible with uniqueness given the provided context, but just to drive home the point, I provide a disjunctive variant of Heim's sage plant sentence (following [10]).

(17) Either Sue didn't buy a^v sage plant,
or she bought eight others along with it$_v$.

It's important to mention at this point that the possibility of anaphora in Partee disjunctions parallels facts concerning presupposition projection. Despite the fact that the a definite description typically presupposes uniqueness, (18) lacks a corresponding uniqueness inference.

(18) Either there isn't a bathroom, or the bathroom is upstairs.

The account of Partee disjunctions which I develop in Sect. 4 leans on this parallel, ultimately unifying (15) and (18) by generalizing the Strong Kleene logic of indeterminacy to a dynamic setting.

3.3 Program Disjunctions

G&S themselves observe that there are cases in which an externally static disjunction makes the wrong predictions. They give the example in (19)—more generally, anaphora from out of a disjunctive sentence is possible when each disjunct contains a parallel indefinite.[8]

(19) A^v professor or an^v assistant professor will attend the meeting of the university board. He$_v$ will report to the faculty.

They use this data to motivate a completely distinct disjunction operator, which they dub *program disjunction*, which is internally static but externally dynamic, and thus captures the data in (19) (although Partee disjunctions are still out of reach). The details won't be important for our purposes, but note that the fact that an alternative, externally static semantics for disjunction is possible in DPL conjures up the same conceptual worry that I've already raised—namely, it's not clear *why* logical expressions manipulate anaphoric information in just the way that they do.

Moreover, once disjunction can be translated into an externally static operator, it's not clear why it only occurs in the kind of instructions instantiated by (19). If disjunction can be externally dynamic, why should anaphora be *impossible* out of a disjunctive sentence elsewhere? Ideally, one would settle on whether the treatment of disjunction is externally static or dynamic. The semantics I'll ultimately end up with will be closer in spirit to G&S's program disjunction. In fact, it turns out that there is a problem with the data motivating G&S's externally static disjunction, which I turn to now.

[8] This observation is often attributed to the later [18]. The intuition behind the analysis is already implied by the fact that the two indefinites are annotates with the same variable.

3.4 Anaphora and Contextual Entailment

G&S's general project involves capturing surface generalizations about anaphora in complex sentences by picking just the right semantics for logical expressions. Rothschild [19] made an observation that shows that this simple picture overlooks the important role of the discourse context. Consider: ordinarily, anaphora out of a disjunctive sentence is impossible, as illustrated by (20).

(20) Either it's a weekday, or a^v critic is watching our play.
#They$_v$ look unhappy.

Rothschild points out that when a witness to the indefinite is subsequently (locally, in this case) contextually entailed, anaphora is possible.

(21) Either it's a weekday, or a^v critic is watching out play.
If it's Saturday today, I want them$_v$ to give us a good review.

Elliott [20] shows that this a very general problem for DPL—other operators, which were thought to be externally static, such as implication, allow for anaphora in similar circumstances; the ultimate suggestion is that a logic which gives connectives an externally dynamic semantics by default is desirable. Later, in Sect. 5, I'll have more to say about how to account for restrictions on anaphora out of disjunction.

Having surveyed some of the most pressing conceptual and empirical issues for DPL,[9] in the next section I begin to develop a new logic for anaphora, building on the Strong Kleene logic of indeterminacy.

4 EDS

EDS stands for *Existential Dynamic Semantics*, or alternatively *Externally-Dynamic Dynamic Semantics*, and it has some signature logical properties which distinguish it from DPL and related theories. I'll explore these properties in more detail later, but briefly:

- Double Negation Elimination is valid in EDS.
- Egli's theorem doesn't hold, but rather a weaker equivalence.
- De Morgan's equivalences hold.
- The logical connectives are a generalization of the Strong Kleene trivalent connectives into a dynamic setting.

[9] For an excellent recent overview of DPL, which expands on many issues which I don't have the space to discuss here, see [21].

4.1 The Basics

At the core of EDS is the idea that pronouns are variables which semantically *presuppose* the existence of an assigned value at a given evaluation point (see especially [22]). This is implemented in the logic formally by emulating partial assignments using an privileged value in the domain of individuals $\#_e$, which corresponds intuitively to the 'unknown' individual. Concretely, assignments are *total* functions from a stock of variables to $D \cup \{\#_e\}$.

In order to simplify the presentation, I'll consider a language with variables and no constants. Since I emulate partiality via the unknown individual, an atomic sentence ϕ receives the obvious (static) trivalent interpretation, where the truth of the atomic sentence at g is unknown just in case the value of any of the variables in the sentence is unknown at g. This is formalized below in (22), where the third truth-value is **unknown**.[10,11]

(22) **Static semantics for atomic sentences**

$$[P(v_1, \ldots, v_n)]^{w,g} = \begin{cases} \textbf{unknown} & g(v_1) = \#_e \ldots \vee \ldots g(v_n) = \#_e \\ \textbf{true} & [P(v_1, \ldots, v_n)]^{w,g} \text{ is not } \textbf{unknown} \\ & \text{and } \langle g(v_1), \ldots, g(v_n)\rangle \in I_w(P) \\ \textbf{false} & [P(v_1, \ldots, v_n)]^{w,g} \text{ is not } \textbf{unknown} \\ & \text{and } \langle g(v_1), \ldots, g(v_n)\rangle \notin I_w(P) \end{cases}$$

There are a number of possibilities for making a DPL-style relational semantics *partial* (see especially [24] for discussion). In EDS, the main innovation is that each of the three truth-values in a trivalent logic corresponds to a DPL-style relational meaning in a dynamic setting, i.e., it keeps track of anaphoric information associated with verification, falsification, and the 'unknown' case in tandem.[12] EDS is therefore a *trivalent* logic; in order to formalize this idea, I recursively define $[\![.]\!]^w_+$, $[\![.]\!]^w_-$, $[\![.]\!]^w_?$, (corresponding to the true, false, and unknown respectively).

(23) **Atomic sentences in EDS**

a. $[\![P(v_1, \ldots, v_n)]\!]^w_+ := \{\, (g, h) \mid g = h \wedge |P(v_1, \ldots, v_n)|^{w,h} \text{ is } \textbf{true} \,\}$

---

[10] Since I'm exclusively concerned with *anaphoric* presuppositions here, I make the simplifying assumption that all predicates are bivalent, i.e., if all of the values of the variables are *known* then an atomic sentence is always either **true** or **false**. One way of extending the logical language in order to model (non-anaphoric) presuppositions while maintaining bivalent predicates would be to incorporate Beaver's unary presupposition operator [23].

[11] Note that there are different ways in which to interpret the third truth value in a trivalent setting, e.g., as standing in for *undefinedness*. Here, it is explicitly referred to as "unknown", since this framing is a natural fit for the Strong Kleene logic of indeterminacy, which is exploited extensively later in the paper. Undefinedness typically goes together with Weak Kleene logic. I'm grateful to an anonymous reviewer for pressing me to clarify this point.

[12] This builds on the dynamic system developed in [25,26], in which outputs are paired with bivalent truth-values.

b. $[\![P(v_1,\ldots,v_n)]\!]^w_- := \{\,(g,h) \mid g = h \wedge |P(v_1,\ldots,v_n)|^{w,h}$ **is false** $\}$
c. $[\![P(v_1,\ldots,v_n)]\!]^w_?$
$:= \{\,(g,h) \mid g = h \wedge |P(v_1,\ldots,v_n)|^{w,h}$ **is unknown** $\}$

Since the logic is trivalent, I give explicit truth and falsity conditions. N.b. that according to (24), a sentence with a free variable v will be **unknown** at g if $g(v) = \#_e$.

(24) **Truth and falsity in EDS**
a. $[\phi]^{w,g}$ **is true** if $\{\,h \mid (g,h) \in [\![\phi]\!]^w_+\,\} \neq \emptyset$
b. $[\phi]^{w,g}$ **is false** if $[\phi]^{w,g}$ is not **true** and $\{\,h \mid (g,h) \in [\![\phi]\!]^w_-\,\} \neq \emptyset$
c. $[\phi]^{w,g}$ **is unknown** otherwise

4.2 Negation

Negation in EDS is a flip-flop operator, defined as in (25) (see also [15]). N.b. that presuppositions project. This is a generalization of Strong Kleene negation, in the sense that each cell in the Strong Kleene truth table is interpreted as a DPL-style relational meaning, as opposed to a truth value.

(25) **Negation in EDS**
a. $[\![\neg\phi]\!]^w_+ := [\![\phi]\!]^w_-$
b. $[\![\neg\phi]\!]^w_- := [\![\phi]\!]^w_+$
c. $[\![\neg\phi]\!]^w_? := [\![\phi]\!]^w_?$

It follows straightforwardly from the flip-flop definition that Double-Negation Elimination is valid:

(26) **Double Negation in EDS:** $\phi \iff \neg\neg\phi$

As I've already discussed, it seems desirable to have a dynamic logic in which (26) holds. The statement of (25) is of course extremely straightforward. In the following I'll show that flip-flop negation makes good predictions in tandem with the other logical operators, once defined.

4.3 Connectives and Embedding Strong Kleene

Now for the logical connectives. I've gestured several times towards the idea that the semantics of the logical connectives is a generalization of Strong Kleene trivalent logic to a dynamic setting. It's now time to make this idea precise. What kind of information does a truth table encode for a binary connective $*$? Well, given the truth values of two sentences ϕ, ψ it tells us how to compute the truth-value of the complex sentence $\phi * \psi$. Each cell in a truth table therefore expresses the result of apply some function from pairs of truth values, to truth values. In a dynamic setting, the values of sentences ϕ, ψ are not truth-values but rather relations. It's therefore natural to interpret each cell in a truth table as specifying a *relational composition*. The classical truth value tells us which polarity the resulting relation belongs to, on the basis of the polarities of the input relations. Exactly how this works will become more readily apparent once I go through some concrete examples, so let's start with the simplest case: conjunction.

In Sect. 1, I give the Strong Kleene 'truth-table' for conjunction in EDS. Just as in Strong Kleene semantics, a conjunctive sentence is only verified if both conjuncts are verified, but here *verification* is interpreted in a dynamic sense—in order to compute the positive extension of the conjunctive sentence, compute the relational composition of the positive extensions of the conjuncts. Falsification is a weaker requirement—there are many different ways in which conjunctive sentences can be falsified in Strong Kleene logic, and in some cases one of the conjuncts is unknown. The negative extension of the conjunctive sentence is the union of all of the dynamic falsifications. The unknown extension is also computed by taking the union of all of the unknown cases, computed dynamically.[13]

$\phi \wedge \psi$	$[\![\psi]\!]^w_+$	$[\![\psi]\!]^w_-$	$[\![\psi]\!]^w_?$
$[\![\phi]\!]^w_+$	$\circ, +$	$\circ, -$	$\circ, ?$
$[\![\phi]\!]^w_-$	$\circ, -$	$\circ, -$	$\circ, -$
$[\![\phi]\!]^w_?$	$\circ, ?$	$\circ, -$	$\circ, ?$

Fig. 1. Strong Kleene conjunction in EDS

EDS has a left-to-right bias directly encoded in the recipe it uses for lifting Strong Kleene semantics into a dynamic setting, since relational composition is non-commutative. Below, I write out the information encoded informally in Fig. 1 as the semantics of conjunction in EDS.[14]

(27) **Conjunction in EDS**

a. $[\![\phi \wedge \psi]\!]^w_+ := [\![\phi]\!]^w_+ \circ [\![\psi]\!]^w_+$

b. $[\![\phi \wedge \psi]\!]^w_- := [\![\phi]\!]^w_- \circ [\![\psi]\!]^w_{+,-,?} \cup [\![\phi]\!]^w_{+,?} \circ [\![\psi]\!]^w_-$

c. $[\![\phi \wedge \psi]\!]^w_? := [\![\phi]\!]^w_+ \circ [\![\psi]\!]^w_? \cup [\![\phi]\!]^w_? \circ [\![\psi]\!]^w_{+,?}$

Before I discuss a concrete application, i.e., modelling discourse anaphora, a remark is in order on the generality of this picture. As an anonymous reviewer points out, the question of what kind of information a truth-table encodes of course generalizes beyond just binary connectives. The picture outlined here can

[13] The generalization of Strong Kleene trivalent semantics to a dynamic setting will out of necessity remain rather impressionistic in this paper. The procedure of lifting truth-functional operators into a dynamic setting has however been made precise in important work by Charlow [26]. Simon Charlow (p.c.) points out that the recipe for lifting the Strong Kleene semantics used here can be formalized as a lifting of the Strong Kleene connectives into the `State.Set` applicative, following [26]. See [20] for more details.

[14] In order to keep the definitions relatively terse, I take advantage of the convention that $[\![\phi]\!]^w_{+,-,?}$ is understood as $[\![\phi]\!]^w_+ \cup [\![\phi]\!]^w_- \cup [\![\phi]\!]^w_?$.

be generalized as follows: one can think of a classical truth-table as encoding a function f from a sequence of n truth-values to a truth-value (where $n > 0$). The polarized dynamic interpretations that EDS deals in can in turn be encoded as a pair consisting of a truth-value and a relation. For each 'cell' in a derived dynamic truth table, I can state a general recipe: it takes as its input f, and a sequence of polarized relation pairs, and gives back a polarized relation; this is formalized below in (28). See [20] for more details.

(28) $$f((t_1, R_1), \ldots, (t_n, R_n)) := \begin{cases} (f(t_1), R_1) & n = 1 \\ (f(t_1, \ldots, t_n), R_1 \circ \ldots \circ R_n) & n > 1 \end{cases}$$

Now let's turn to a concrete appliation of the EDS semantics for conjunction: discourse anaphora. First let's define discourse referent introduction in EDS. Since negation is a flip-flop operator in EDS, it's important to ensure that a negated existential statement doesn't introduce anaphoric information, while preserving DNE. This is accomplished by syncategorematically defining existential quantification in terms of (a) conjunction (27), (b) DPL-style random assignment (29), and (c) a 'positive closure' operator. Positive closure simply ensures that its negative extension is always a test (i.e., anaphorically inert).

(29) **Random assigment in EDS**

a. $[\![\varepsilon_v]\!]^w_+ := \{ (g,h) \mid g[v]h \}$

b. $[\![\varepsilon_v]\!]^w_- := \emptyset$

c. $[\![\varepsilon_v]\!]^w_? := \emptyset$

(30) **Positive closure in EDS**

a. $[\![\dagger\phi]\!]^w_+ := [\![\phi]\!]^w_+$

b. $[\![\dagger\phi]\!]^w_- := \{ (g,h) \mid g = h \wedge [\phi]^{w,g} \text{ is } \mathbf{false} \}$

c. $[\![\dagger\phi]\!]^w_? := [\![\phi]\!]^w_?$

Existential quantification is defined syncategorematically, just as in DPL but with the addition of $\dagger$.

(31) **Existential quantification in EDS**

$\exists_v\phi := \dagger(\varepsilon_v \wedge \phi)$

I'll now establish some useful facts relating to the treatment of discourse anaphora in EDS. Note that $[\![\phi \wedge \psi]\!]^w_+$ is a simple relational composition. Consequently, concentrating just on the positive extension, associativity holds (32), and therefore the account of discourse anaphora from DPL is maintained. This is easy to see, since the positive extension of random assignment is the same as DPL random assignment, and positive closure is vacuous with respect to positive extensions.

(32) **Positive associativity of conjunction in EDS:**

$[\![(\phi \wedge (\psi \wedge \sigma))]\!]^w_+ = [\![(\phi \wedge \psi) \wedge \sigma]\!]^w_+$

The interaction between negation and discourse referent introduction is one respect in which EDS substantially departs from DPL. In DPL, negative existential statements are anaphorically inert by dint of the special properties of negation. In EDS, conversely, negation is more classical—DNE is valid—and negated existential statements are anaphorically inert by dint of the special properties of *positive closure*, which ensures that an existential statement is a *negative test* (i.e., its negative extension is a test). This is illustrated in (33).

(33) **Negative existential statements are negative tests:**
$[\![\exists_v P(v)]\!]^w_-$
a. $= [\![\dagger(\varepsilon_v \wedge P(v))]\!]^w_-$
b. $= \{ (g,h) \mid g = h \wedge [\varepsilon_v \wedge P(v)]^{w,g} \text{ is } \textbf{false} \}$
c. $= \{ (g,h) \mid g = h \wedge I_w(P) = \emptyset \}$

There's more to be said about the negative extension of conjunctive sentences, where (as I'll show), one observes failures of associativity. First though, I'll discuss the semantics of disjunction on EDS, illustrating how it resolves the vexing problem of Partee disjunctions.

4.4 Disjunction

Just as with conjunction, the semantics of disjunction in EDS is a lifting of the Strong Kleene trivalent semantics into a dynamic setting. This is illustrated in Fig. 2. With conjunction, there was essentially one way of dynamically verifying the sentence, but many ways of dynamically falsifying. With disjunction, the situation is the reverse: there are many ways of dynamically verifying, but only one way of dynamically falsifying.

$\phi \vee \psi$	$[\![\psi]\!]^w_+$	$[\![\psi]\!]^w_-$	$[\![\psi]\!]^w_?$
$[\![\phi]\!]^w_+$	$\circ, +$	$\circ, +$	$\circ, +$
$[\![\phi]\!]^w_-$	$\circ, +$	$\circ, -$	$\circ, ?$
$[\![\phi]\!]^w_?$	$\circ, +$	$\circ, ?$	$\circ, ?$

Fig. 2. Strong Kleene disjunction in EDS

The Strong Kleene truth-table in Fig. 2, where each cell is interpreted as a relational composition, corresponds to the EDS semantics of disjunction laid out below.

(34) **Disjunction in EDS**
a. $[\![\phi \vee \psi]\!]^w_+ := [\![\phi]\!]^w_+ \circ [\![\psi]\!]^w_{+,-,?} \cup [\![\phi]\!]^w_{-,?} \circ [\![\psi]\!]^w_+$
b. $[\![\phi \vee \psi]\!]^w_- := [\![\phi]\!]^w_- \circ [\![\psi]\!]^w_-$
c. $[\![\phi \vee \psi]\!]^w_? := [\![\phi]\!]^w_- \circ [\![\psi]\!]^w_? \cup [\![\phi]\!]^w_? \circ [\![\psi]\!]^w_{-,?}$

The crucial insight which will underlie the account of Partee disjunctions in EDS is that, one way of dynamically verifying a disjunctive sentence is by composing the negative extension of the first disjunct with the positive extension of the second. In EDS, since DNE is valid, a negative extension can introduce a discourse referent. In order to go through how this works, Il work through a simple example (35).

(35) Either there's no^v bathroom, or it's upstairs.
$\neg\exists_v B(v) \vee U(v)$

First, let's spell out the negative and positive extensions of the first disjunct; the positive extension tests whether there are no bathrooms (thanks to positive closure), and the negative extension introduces a bathroom discourse referent.

(36) $[\![\neg\exists_v B(v)]\!]^w_+ = [\![\exists_v B(v)]\!]^w_- = \{\,(g,h) \mid g = h \wedge I_w(B) = \emptyset\,\}$

(37) $[\![\neg\exists_v B(v)]\!]^w_- = [\![\exists_v B(v)]\!]^w_+ = \{\,(g,h) \mid g[v]h \wedge h(v) \in I_w(B)\,\}$

The second disjunct is an open sentence, so it has a standard trivalent test semantics. In order to compute the positive extension of the disjunctive sentence, I consider all ways of dynamically verifying the disjunction.

- One salient possibility is that one verifies the disjunction by falsifying the first disjunct, and verifying the second disjunct. Falsifying the first disjunct introduces a bathroom discourse referent which is dynamically retrieved when verifying the second disjunct (38).
- Another way of verifying the disjunction is by verifying the first disjunct, in which case the second disjunct is irrelevant—this is captured in Strong Kleene semantics, by taking the relational composition with the positive/negative/unknown extension of the second disjunct. Since the second disjunct is a test, this is equivalent to the positive extension of the first disjunct (39).
- Finally, I union everything together in (40).

(38) $[\![\neg\exists_v B(v)]\!]^w_- \circ [\![U(v)]\!]^w_+$
$= [\![\exists_v B(v)]\!]^w_+ \circ [\![U(v)]\!]^w_+$
$= \{\,(g,h) \mid g[v]h \wedge h_v \in I_w(B) \wedge h_v \in I_w(U)\,\}$

(39) $[\![\neg\exists_v B(v)]\!]^w_+ \circ [\![U(v)]\!]^w_{+,-,?} = \{\,(g,h) \mid g = h \wedge I_w(B) = \emptyset\,\}$

(40) $[\![\neg\exists_v B(v) \vee U(v)]\!]^w_+ = \{\,(g,h) \mid g[v]h \wedge h_v \in I_w(B) \wedge h_v \in I_w(U)\,\} \cup \{\,(g,h) \mid g = h \wedge I_w(B) = \emptyset\,\}$

This captures the attested *existential* truth-conditions of Partee disjunctions, which I argued for in Sect. 3.2; the positive extension of the Partee disjunction will be non-empty if, either: (a) there is a bathroom upstair (in which case, introduce a bathroom upstairs discourse referent), or (b) (there is no bathroom).

This is (arguably) a desirable result! There is however a pressing issue that arises under the EDS semantics for disjunction, which underlies an apparent issue for the semantics more generally. Namely, Partee disjunctions *conditionally* introduce discourse referents. More generally, our semantics for disjunction is *externally dynamic* in the sense of [3]. This seems, on the face of it, incompatible with the evidence that disjunction is externally static, as discussed way back in Sect. 2.

Manifestations of this problem can be seen elsewhere. For example, although I won't discuss this in detail, de Morgan's equivalences are valid in EDS.[15] One consequence of this is that $\neg\exists_v B(x) \vee U(x)$ is equivalent to $\neg(\exists_v B(v) \wedge \neg U(v))$ via de Morgan's and DNE. This means that negated conjunctions can conditionally introduce discourse referents too. Another equivalent sentence in EDS[16] is $\exists_v B(x) \rightarrow U(x)$—similarly, G&S argue that material implication is externally static, but in EDS the implicational sentence conditionally introduces a discourse referent.

In the next section, I'll show that, far from being a fatal problem, making external dynamicity the ordinary case is a desirable feature for a dynamic logic. I've already provided some empirical evidence for this in the form of program disjunctions, discussed in Sect. 3.3, and Rothschild's observation, discussed in Sect. 3.4. EDS will capture both of these datapoints, while maintaining a certain degree of restrictiveness, once integrated into a theory of discourse pragmatics.

---

[15] I'll simply note here that the validity of de Morgan's in the presence of anaphoric dependencies seems independently desirable given our intuitions about natural language. The following sentences are all arguably truth-conditionally equivalent.

(1) a. Either there's no bathroom, or it's upstairs.
 b. It's not the case that there's a bathroom and it's not upstairs.
 c. If there's a bathroom, then it's upstairs.

[16] Assuming a Strong Kleene semantics for material implication. Something interesting to note here is that EDS predicts existential truth conditions for donkey sentences, unlike, e.g., [3]. Egli's corrolary therefore doesn't hold.

This is by no means a bad prediction—it has been widely reported that such existential readings are attested for donkey sentences, as alluded to in fn. 7 (see also [17,27]). The following example, for example, is clearly true if Gabor owns two credit cards but only pays with one of them.

(1) If Gabor has a^x credit card, he'll pay with it$_x$.

The empirical picture is however much more complicated, and donkey sentences do often have stronger, universal readings. Relatedly, [15] reports that Partee disjunctions have universal readings—unlike donkey sentences, to my knowledge very little work has been done examining the distribution of existential and universal readings of Partee disjunctions. A detailed discussion of universal readings will have to wait for another occasion, but see [28] for one recent approach.

5 Discourse Pragmatics

5.1 Update

In order to give an account of the dynamics of disjunction, it's important to understand how discourse referents are introduces in context. In certain dynamic theories, such as Heim's *File Change Semantics* [2], the relationship between the semantic value of ϕ and what it means to assert ϕ is almost trivial, since on such theories sentences themselves denote updates on information states (see also [29]). Since EDS is a relational theory, much like DPL, as well as encoding partiality, I need to state a concrete bridge principle in order to integrate EDS with a Heimian notion of information states. Update in EDS is defined as in (41).

(41) **Update in EDS:**

$$c[\phi] = \begin{cases} \bigcup_{(w,g)\in c} \{ (w,h) \mid (g,h) \in [\![\phi]\!]^w_+ \} & \forall (w,g) \in c \begin{bmatrix} [\phi]^{w,g} \textbf{ is true} \\ \text{or } [\phi]^{w,g} \text{ is } \textbf{false} \end{bmatrix} \\ \text{undefined} & \text{otherwise} \end{cases}$$

Here, I take information states to be sets of world-assignment pairs [2]. 'Initial' states (i.e., those where no discourse referents have been introduced) are those paired with the unique assignment which maps every variable to $\#_e$ (I'll write the initial assignment as $[]$). Updating an information state c with a sentence ϕ is *defined* just in case ϕ is *contextually bivalent.*[17] If defined, the updated information state is computed by gathering up, at each evaluation point $i \in c$, the positive extension of ϕ at i. EDS sentences therefore update information states by (i) eliminating worldly possibilities, and (ii) introducing discourse referents, i.e., expanding anaphoric possibilities.

An immediate consequence of the notion of update in (41) is that an open sentence $P(v)$ presupposes at c that v is 'defined' at every evaluation point $i \in c$. This is exactly the notion of *familiarity* introduced by Heim [2,31], but here derived from a partial DPL-like dynamic semantics plus a generalization of Stalnaker's bridge. To be precise: *definedness* is a condition placed on individual evaluation points, whereas *familiarity* is a (derivative) universal condition placed on the entire input context.

5.2 Disjunction and Contingency

Disjunctive assertions in natural language are subject to a contingency requirement (43). I state this formally as a felicity condition on assertion in (43), making use of a notion of *worldly content* defined in (42)—the idea here is just that it's possible to retrieve the 'classical' Stalnakerian content of a Heimian

[17] This is what von Fintel calls 'Stalnaker's bridge' [30], in the context of a dynamic setting.

information state. This captures the intuition that a disjunctive sentence cannot be felicitously asserted if one of the disjuncts is contextually trivial.[18]

(42) **Worldly content:** $W(c) := \{ w \mid \exists g[(w, g) \in c] \}$

(43) **Contingency requirement:**
Assertion of a sentence of the form $\phi \vee \psi$ is felicitous in c iff $\mathbf{W}(c[\phi])$ and $\mathbf{W}(c[\psi])$ are non-empty proper subsets of $\mathbf{W}(c)$.

The update rule in (41), together with the contingency requirement in (43) accounts for G&S's observations concerning the apparent external staticity of disjunction, as well as Rothschild's observation, discussed in Sect. 3.4. To see why, consider the simple example in (44). The first disjunct $\exists_v P(x)$ is contextually trivial at c, unless some worlds in c are worlds s.t., $I_w(P) = \emptyset$. This guarantees that, so long as (43) is satisfied, updating an information state with (44) will result in an updated information state containing at least some non-P worlds, where discourse referents aren't introduced. This means that a subsequent open sentence such as $Q(v)$ cannot be felicitously asserted, since (44) can't make v *familiar*.

(44) $\exists_v P(v) \vee Q(a)$

Crucially, if the non-P worlds are subsequently eliminated, v might become familiar later in the discourse, for example if an assertion is made that contextually entails the first disjunct. In this case, anaphora will be possible since familiarity will be satisifed.

This general explanatory strategy can be extended to other apparently cases of external staticity, once the contingency requirement in (43) is generalized to other complex sentences. For example, assertion of sentences of the form $\neg(\phi \wedge \psi)$ typically requires that $\neg\phi$ and $\neg\psi$ are not contextually trivial.

One interesting thing to note is that the requirement as stated in (43) doesn't quite work as stated for Partee disjunctions, since it doesn't take into account the possibility of an anaphoric dependency between disjuncts. I address this issue in detail in [32].

5.3 Program Disjunctions

I'm now in a position to explain why program disjunctions are an apparent exception to the more general properties of disjunctive assertions in discourse. Following Groenendijk and Stokhof's DPL account [3], I assume that what makes program disjunctions special is that each disjunct is an existential statement

[18] Various pragmatic justifications can be given for the formal contingency requirement stated in (43). What is important for my purposes is that if a disjunctive sentence is asserted by a speaker s in a context which trivializes one of the disjuncts, the assertion is judged to be 'odd'.

introducing a discourse referent at the same variable. A schematic case is provided in (45). *Unlike* DPL, EDS accounts for the behavior of program disjunctions *without* having to assume that disjunction is ambiguous between externally static and externally dynamic variants.

(45) $\exists_v P(v) \vee \exists_v Q(v)$

The contingency requirement insists that there be some P worlds, and some non-P worlds in c for (45) to be assertable, as well as some Q-worlds, and some non-Q worlds. Once (45) is asserted however, all non-P, non-Q worlds will be eliminated. This leaves only P-worlds and Q-worlds, each of which is associated with a discourse referent at v. (45) therefore makes v familiar, and subsequent anaphora is (accurately) predicted to be possible by EDS.[19]

5.4 Internal Staticity

There is a loose end from Sect. 3 that I have yet to address in the more permissive setting of EDS—namely, why is disjunction internally static? The problematic data is given below.

(46) #Either there's a^v bathroom, or it$_v$'s upstairs.

In fact, in order to capture Partee disjunctions, it seems essential to allow for anaphoric information to pass between disjuncts, so (46) seems to constitute something of a mystery. In fact, the infelicity of (46) in a context where v isn't familiar is expected on the basis of the contingency requirement. Consider the LF of (46):

(47) $\exists_v B(v) \vee U(v)$

If the first disjunct is true, the second is contextually bivalent, but if the first disjunct is false, the truth of the second disjunct is partial, and dependent on the input assignment. Eliding the full computation, the positive and negative extensions of (47) is given below:

(48) $[\![.]\!]^w_+ = \{\, (g,h) \mid g[v]h \wedge g(v) \in I_w(B) \,\} \cup \{\, (g,h) \mid g = h \wedge I_w(B) = \emptyset \wedge h(v) \in I_w(U) \,\}$

(49) $[\![.]\!]^w_- = \{\, (g,h) \mid g = h \wedge I_w(B) = \emptyset \wedge h(v) \notin I_w(U) \,\}$

[19] There seem to be information-structural constraints on program disjunctions in natural language which are beyond the remit of EDS. For example, it seems that some degree of parallelism is required to hold between the disjuncts. Singular anaphora, by my reckoning, is extremely difficult in the following example:

(1) Either a^v linguist sneezed, or the meeting was interrupted by a^v philosopher. ?She was very rude.

I speculate that this is related to constraints on co-indexing. I leave this interesting issue to future work on program disjunctions.

For the disjunctive sentence to be assertable, every $(w, g) \in c$ should be such that either w is a B-world, or w is a non-B world and $g(v)$ is defined. Given the contingency requirement then (suitably generalized to allow for anaphoric dependencies), there should be non-B parts of c, in which case (47) requires a familiar discourse referent v in order to be assertable.

In fact, the empirical picture is potentially even more nuanced than this. Filipe Hisao Kobayashi (p.c.) observes that anaphora seems to be possible in (50).

(50) Either there's a^v bathroom upstairs, or it$_v$'s downstairs.

Here I'll tentatively suggest that the contrast between examples like (46) and (50) is due to the different Logical Forms available to existential statements. Concretely, an existential statement in natural language can be translated either as discourse anaphora (51a) or as a existentially-quantified formula (51b). In a theory such as DPL, (51a) and (51b) are equivalent (Egli's theorem). In EDS on the other hand, (51a) and (51b) are positively equivalent but negatively distinct, due to the fact that conjunction isn't associative.[20]

(51) a. $\exists_v B(v) \wedge U(v)$
b. $\exists_v (B(v) \wedge U(v))$

Concretely, the negative extension of (51b) is always a test, due to positive closure taking widest scope. The negative extension of (51a) on the other hand conditionally introduces a discourse referent. In (50), I conjecture, the first disjunct is translated as in (51a). I leave a more detailed assessment of examples such as (50) to future research.

6 Comparison to Alternatives

Although it will be impossible to provide a detailed comparison between EDS and related proposals, some parallels and correspondences are worth mentioning.

The semantics of existential quantification in EDS—decomposed into positive closure, conjunction, and random assignment, is closely related to the system developed in Mandelkern's work [33]. Mandelkern develops a logic of anaphora which is bivalent and classical, but supplemented with an extra dimension of meaning—*witness bounds*. The witness bounds of an existential statement ensure that a discourse referent is conditionally introduced if there is a witness to the existential statement, and thereby maintains external staticity of a negated existential statement while validating DNE. The workings of witness bounds are highly reminiscent of my positive closure operator, and there are other compelling logical correspondences between my theory of Mandelkern's which

[20] This kind of suggestion raises the issue of how exactly natural language sentences can be mapped compositionally to EDS Logical Forms. After all, one doesn't want to allow for too much flexibility, otherwise the resulting grammar won't be sufficiently constrained.

deserve further exploration. One respect however in which the theories diverge is that, in EDS, the familiarity requirement associated with a pronoun/free variable is just an ordinary presupposition. In Mandelkern's theory, the correspondence between presupposition projection and anaphoric accessibility is not straightforwardly captured.

Hofmann [34,35] tackles many of the same problems discussed here within the context of a much more expressive system based on CDRT [36] and intensionalized discourse referents [37]. An appealing property of Hofmann's system is that it can handle modality and modal subordination. This is important in accounting for certain cases of anaphora from out of a negative sentence, such as (52).

(52) Colin doesn't own a^v car, but it$_v$ would be a Subaru.

It remains to be seen to what extent Hofmann's insights can be incorporated into EDS, in order to expand its empirical remit.

Finally, it would be remiss of me not to mention the connection between EDS and earlier work by Rothschild [19], which also attempts to account for patterns of anaphoric accessibility using the trivalent account presupposition projection, and which thereby consistitutes an important precursor to EDS. There are a couple of important differences between EDS and Rothschild's proposal—here, a *left-to-right* bias arises due to the way in which the Strong Kleene connectives are lifted into a dynamic setting (i.e., using relational composition). On Rothschild's account, the trivalent logic itself must be given a left-to-right bias [38–40] in order to account for linear asymmetries in anaphora. Furthermore, in order to account for, e.g., Partee disjunctions, Rothschild stipulates that classically transparent material may be freely inserted into Logical Forms. This mechanism is somewhat ad-hoc and leads to concerns of over-generation. EDS consistutes a clear improvement, in the sense that Partee disjunctions are follow from standard dynamic mechanisms for capturing cross-sentential anaphora.

7 Conclusion and Outlook

In this paper, I've sketched a new kind of dynamic logic: EDS. EDS incorporates the insights of Groenendijk & Stokhof's *Dynamic Predicate Logic*, and trivalent approaches to presupposition projection. A core tenet of EDS is that the dynamics of the logical connectives should not be stipulated, but rather arise as a generalization of the Strong Kleene connectives into a dynamic setting. This approach is conceptually appealing, as it maintains a certain degree of predictiveness while establishing a tight connection between patterns of anaphoric accessibility and presupposition projection, following, e.g., [19].

EDS is *more classical* than orthodox logics of anaphora such as DPL in important respects—for example, DNE is valid. This is an important result, as empirical evidence suggests that classical equivalences such as DNE and de Morgan's don't break down in the presence of anaphoric dependencies. There are also striking respects in which EDS differs from DPL. To recap, neither Egli's theorem nor Egli's corrolary hold in EDS. This is surprising, since Egli's theorem is often framed as *the* central logical property of dynamic theories. Instead, a weaker variant of Egli's theorem holds, just with respect to *positive* extensions. Another major departure is that EDS predicts existential readings across the board, including for donkey sentences.

Much work remains to be done in investigating the inferential properties of EDS, and extending the central ideas outlined here to a broader empirical domain, encompassing quantification, plurality, and modality.

Acknowledgments. Aspects of this work have been presented in various venues, including at Rutgers, NYU, MIT, ENS, and most recently at the Third Tsinghua Interdisciplinary Workshop on Logic, Language, and Meaning. I'm grateful to participants on all such occasions for insightful and challenging feedback on this material, which has shaped the current form. The logic outlined in this paper was developed for the Spring 2022 *Topics in Semantics* seminar at MIT, and I'm especially grateful to Filipe Hisao Kobayashi and Enrico Flor for their input. I remain solely responsible for any mistakes.

References

1. Karttunen, L.: Discourse referents. In: McCawley, J.D. (ed.) Syntax and Semantics, vol. 7, pp. 363–386. Academic Press (1976)
2. Heim, I.: The Semantics of Definite and Indefinite Noun Phrases. PhD thesis, University of Massachusetts - Amherst (1982)
3. Groenendijk, J., Stokhof, M.: Dynamic predicate logic. Linguist. Philos. **14**(1), 39–100 (1991)
4. Simons, M.: Disjunction and Anaphora. Semantics Linguist. Theory **6**, 245–260 (1996). https://doi.org/10.3765/salt.v6i0.2760
5. Heim, I.: On the projection problem for presuppositions. In: Proceedings of WCCFL 2, pp. 114–125, Stanford University (1983)
6. Soames, S.: Presupposition. In: Gabbay, D., Guenthner, F. (eds.) Handbook of Philosophical Logic: Topics in the Philosophy of Language, vol. 4. pp. 553–616. Springer, Netherlands, Dordrecht (1989). https://doi.org/10.1007/978-94-009-1171-0_9
7. Schlenker, P.: Be articulate: A Pragmatic Theory of Presupposition Projection. Theoretical Linguist. **34**(3) (2008). https://doi.org/10.1515/THLI.2008.013
8. Schlenker, P.: Local contexts. In: Semantics and Pragmatics, vol. 2 (2009). https://doi.org/10.3765/sp.2.3
9. Schlenker, P.: Local contexts and local meanings. Philoso. Stud. Int. J. Philos. Analytic Tradition **151**(1), 115–142 (2010). https://doi.org/10.2307/40856594
10. Mandelkern, M., Rothschild, D.: Definiteness projection. Nat. Lang. Seman. **28**(2), 77–109 (2019). https://doi.org/10.1007/s11050-019-09159-2
11. Heim, I.: E-type pronouns and donkey anaphora. Linguist. Philos. **13**(2), 137–177 (1990). https://doi.org/10.1007/BF00630732
12. Elbourne, P.: Situations and Individuals. PhD thesis, Massachussetts Institute of Technology (2005)
13. Elbourne, P.: Definite Descriptions. Oxford University Press, Oxford (2013)
14. Mandelkern, M., Rothschild, D.: Independence day? J. Semant. **36**(2), 193–210 (2019). https://doi.org/10.1093/jos/ffy013

15. Krahmer, E., Muskens, R.: Negation and disjunction in discourse representation theory. J. Semant. **12**(4), 357–376 (1995). https://doi.org/10.1093/jos/12.4.357
16. Gotham, M.: Double negation, excluded middle and accessibility in dynamic semantics. In: Schlöder, J.J., McHugh, D., Roelofsen, F. (eds.) Proceedings of the 22nd Amsterdam Colloquium, pp. 142–151 (2019)
17. Chierchia, G.: Dynamics of Meaning - Anaphora, Presupposition, and the Theory of Grammar. University of Chicago Press, Chicago (1995)
18. Stone, M.D.: 'Or' and Anaphora. Semantics Linguist. Theory **2**, 367–386 (1992). https://doi.org/10.3765/salt.v2i0.3037
19. Rothschild, D.: A trivalent approach to anaphora and presupposition. In: Cremers, A., van Gessel, T., Roelofsen, F. (eds.) Proceedings of the 21st Amsterdam Colloquium, pp. 1–13 (2017)
20. Elliott, P.D.: Towards a Principled Logic of Anaphora. Unpublished manuscript (2020)
21. Gillies, A.S.: On groenendijk and stokhof's "dynamic predicate logic". In: McNally, L., Szabó, Z.G. (eds.) A Reader's Guide to Classic Papers in Formal Semantics, pp. 121–153. Springer International Publishing, Cham (2022). https://doi.org/10.1007/978-3-030-85308-2_8
22. Van den Berg, M.: Full Dynamic plural logic. In: Proceedings of the Fourth Symposium on Logic and Language (1996)
23. Beaver, D.I.: Presupposition and Assertion in Dynamic Semantics. CSLI Publications (2001)
24. Van den Berg, M.H.: Some Aspects of the Internal Structure of Discourse. The Dynamics of Nominal Anaphora (1996)
25. Charlow, S.: On the Semantics of Exceptional Scope. PhD thesis, Rutgers University, New Brunswick (2014)
26. Charlow, S.: Static and dynamic exceptional scope. To Appear J. Semant. (2020)
27. Kanazawa, M.: Weak vs. strong readings of donkey sentences and monotonicity inference in a dynamic setting. Linguist. Philos. **17**(2), 109–158 (1994)
28. Champollion, L., Bumford, D., Henderson, R.: Donkeys under Discussion. Semant. Pragmatics **12**, 1 (2019). https://doi.org/10.3765/sp.12.1
29. Veltman, F.: Defaults in update semantics. J. Philos. Log. **25**(3), 221–261 (1996)
30. Von Fintel, K.: What is presupposition accommodation, again?*. Philos. Perspect. **22**(1), 137–170 (2008). https://doi.org/10.1111/j.1520-8583.2008.00144.x
31. Heim, I.: File change semantics and the familiarity theory of definiteness. In: Meaning, Use, and Interpretation of Language, pp. 164–189. De Gruyter (1983). https://doi.org/10.1515/9783110852820.164
32. Elliott, P.D.: Partee Conjunctions and Free Choice with Anaphora. Handout from a talk given at LFRG, MIT (2022)
33. Mandelkern, M.: Witnesses. Linguist. Philos. (2022)
34. Hofmann, L.: The anaphoric potential of indefinites under negation and disjunction. In: Schlöder, J.J., McHugh, D., and Roelofsen, F. (eds.) Proceedings of the 22nd Amsterdam Colloquium, pp. 181–190 (2019)
35. Hofmann, L.: Anaphora and Negation. PhD thesis, University of California Santa Cruz (2022)
36. Muskens, R.: Combining montague semantics and discourse representation. Linguist. Philos. **19**(2), 143–186 (1996). https://doi.org/10.1007/BF00635836
37. Stone, M.: Reference to Possible Worlds. Technical Report 49, Rutgers University Center for Cognitive Science (1999)
38. Peters, S.: A truth-conditional formulation of karttunen's account of presupposition. Synthese **40**(2), 301–316 (1979). https://doi.org/10.1007/BF00485682

39. George, B.R.: Predicting Presupposition Projection - Some Alternatives in the Strong Kleene Tradition
40. George, B.R.: A New Predictive Theory of Presupposition Projection. In: Proceedings of SALT 18, pp. 358–375. Cornell University, Ithaca, NY. (2008)

Scope Reconstruction and Interpretation of Copies

Paul Law(✉)

The Chinese University of Hong Kong, Shatin, Hong Kong
paulslaw@cuhk.edu.hk

Abstract. We consider the relative scope of quantifiers, modals and negation in A-movement. We present novel data suggesting that the apparent narrow scope of the quantifiers with respect to the modal is due its logical relation with the wide scope reading. We give the same account for the lack of narrow scope reading with respect to negation. We claim that the narrow scope reading is possible when it is logically implied by the wide scope reading. We argue that late merger in extraposition has no effect on quantifier scope. We resolve certain problems arising from the interactions between scope and binding by differentiating lexical features from morphosyntactic features. These are interpreted not all at the same time but dynamically according to their relevance to the grammatical principles.

Keywords: A-movement · Copy theory · Syntax-semantics interface

1 Introduction

Reconstruction is a phenomenon in which a displaced phrase behaves as if it is in its original position. In (1a), the anaphor *himself* contained in the object can be bound by the subject *John*, but that in the subject in (1b) cannot be bound by the object *John*:

(1) a. John told many stories about himself.
b. *Many stories about himself dumbfounded John.
c. How many stories about himself did John tell?
d. How many stories about himself did John tell <how many stories about himself>?

The reason for the contrast, simplified for our purposes here, is that the antecedent *John* c-commands and therefore binds the anaphor in (1a), but does not do so in (1b). However, if the object is fronted to a yet higher position than the subject as in (1c), the anaphor contained in it can be bound by the subject. This phenomenon has been extensively studied for decades [1, 9–13, 15, 20–23, 26, 27, 29–31]. A recent popular account is the copy theory of movement [5], according to which a displaced phrase leaves behind a copy of itself in the position from which it moves, represented in the angle-brackets, as in (1d). When the copy in the original position is interpreted, the structure is exactly like the one in (1a) where the anaphor is bound by the antecedent *John*.

D. Deng et al. (Eds.): TLLM 2022, LNCS 13524, pp. 99–115, 2023.
https://doi.org/10.1007/978-3-031-25894-7_5

The same account can be given to the raising example in (2a), an instance of A-movement in contrast with A-bar-movement in (1c), to explain the scope ambiguity of the sentence where the subject *some senator* may have either wide or narrow scope with respect to the modal *likely*, indicated in (3b,c) respectively:

(2) a. Some senator is likely to support the motion.
b. $\exists x(\textbf{likely}(x\ \textbf{support the motion}))$ $\exists > likely$
c. $\textbf{likely}(\exists x(x\ \textbf{support the motion}))$ $likely > \exists$
d. Some senator is likely [<some senator> to support the motion]

The syntactic representation of (2a) is given in (2d) where the displaced phrase *some senator* leaves a copy of itself when it moves to the matrix clause. The wide scope reading of *some senator* is the result of interpreting the upper copy, and the narrow scope reading arises from interpreting the lower copy. The latter reading is commonly known as **scope reconstruction**.

In this paper, we consider certain facts regarding the interactions between scope reconstruction on the one hand, and negation and anaphoric binding on the other. The interest of these facts is that they lead to a more refined understanding of scope and how the formal features in the copies are to be interpreted. We claim that in contrast with A-bar-movement, only the highest copy of an A-moved phrase is interpreted (see Sect. 6 for some qualifications). This amounts to the same idea that A-movement does not reconstruct [9, 19] (Sect. 2). We argue that the narrow scope reading of the quantifier in cases like (2a) is not due to scope reconstruction, but is due to the logical relation between it and the wide scope reading (Sect. 3). Our analysis strengthens the view that A-movement does not reconstruct (Sect. 4). We further show that late merger of adjuncts in extraposition has no effect on quantifier scope (Sect. 5). We suggest that formal features be differentiated as lexical features and morphosyntactic features. They are not interpreted all at the same time, but are selectively interpreted according to their being relevant to particular principles of grammar (Sect. 6).

2 Scope Reconstruction and Negation

Although the narrow scope reading of the quantifier in (2a) can be accounted for by interpreting the lower copy, the same interpretation would lead to unexpected results.

The example in (3a), differing from that in (2a) in the presence of negation in the matrix clause, does not allow for a narrow scope reading of the quantifier:[1]

(3) a. Some senator is not likely to support the motion.
b. $\exists x(\neg(\textbf{likely}(x\ \textbf{support the motion})))$ $\exists > \neg > likely$
c. $\neg(\textbf{likely}(\exists x(x\ \textbf{support the motion})))$ $^*\neg > likely > \exists$
d. Some senator is not likely [<some senator> to support the motion]

[1] Examples like (3a) with negation in the matrix clause contrast with those with negation in the embedded clause that [6, 19] argue to be evidence against A-movement reconstruction (see Sect. 4).

The difference between the reading in (3b) and the reading in (3c) is that the former remains true in the case where some other senator is likely to support the motion, while the latter denies the likelihood that some senator or other will support it.

The same can be said of the example in (4a). Without negation, both wide and narrow scope readings are available:

(4) a. At least ten senators are likely to support the motion.
   b. **at-least-ten** ***x*****, senator(*****x*****) (likely(*****x*** **support the motion))**
      at least ten senators > likely
   c. **likely(at least ten** ***x*****, senator(*****x*****) (*****x*** **support the motion))**
      likely > at least ten senators
   d. At least ten senators are likely [<at least ten senators> to support the motion]

The scope ambiguity in (4a) can be accounted for in exactly the same way as that in (2a).

However, as in (3a), the narrow scope reading of the quantifier disappears when negation is present in the matrix clause. The example in (5a) only has the reading in which the quantifier has wide scope over both negation and *likely*:

(5) a. At least ten senators are not likely to support the motion.
   b. **at-least-ten** ***x*****, senator(*****x*****) (¬ likely(*****x*** **supports the motion)))**
      at least ten senators > not > likely
   c. **¬ likely(at-least-ten** ***x*****, senator(*****x*****) (*****x*** **supports the motion)))**
      not > likely > at least ten senators
   d. At least ten senators are not likely [<at least ten senators> to support the motion]

The wide scope reading of the quantifier in (5b) is true in a situation where at least ten senators are not likely to support the motion but at least ten others are likely to support it. The narrow scope of the quantifier in (5c) is false in this situation.

The facts in (3a) and (5a) thus show that negation blocks scope reconstruction. If the lower copy in the structures in (3d) and (5d) were to be interpreted, then these sentences should have the narrow reading of the quantifier, contrary to fact. It thus must be that the lower copy of the quantifier is not interpreted in these cases, amounting to the same idea that A-movement mostly does not reconstruct ([5, 6, 20], though based on some different facts [2, 9–12, 16, 19, 30], among many others). If so, then we must ask how the narrow scope readings of the quantifiers in (2a) and (4a) arise, specifically, whether they come about as a result of the lower copies being interpretable in the absence of negation, or whether they arise in some other way.

It seems difficult to entertain the first view, for there is apparently no natural way to explain why negation in the matrix clause should make it impossible to interpret the copy in the embedded clause. We suggest that generally in A-movement only the highest copy be interpreted (see Sect. 4 for some qualifications); the lack of the narrow scope reading in (3a) and (5a) is thus accounted for. This in turn means that we need to explain how the sentences in (2a) and (4a) may have the narrow scope readings of the quantifiers.

For the sentences in (2a) and (4a), repeated in (6a) and (7a) respectively, it is clear that their wide scope readings of the quantifiers logically imply their narrow scope readings in (6b) and (7b) respectively:

(6) a. Some senator is likely to support the motion. (=(2a))
b. $\rightarrow$ It is likely that some senator supports the motion.

(7) a. At least ten senators are likely to support the motion. (=(4a))
b. $\rightarrow$ It is likely that at least ten senators support the motion.

That is, when the wide scope readings of the quantifiers are true, the narrow scope readings are also true.

For the sentences in (8a) and (9a) with negation in the matrix clause, however, the wide scope readings of the quantifiers do not logically imply the narrow scope readings:

(8) a. Some senator is not likely to support the motion. (=(3a))
b. $\nrightarrow$ It is not likely that some senator supports the motion.

(9) a. At least ten senators are not likely to support the motion. (=(5a))
b. $\nrightarrow$ It is not likely that at least ten senators support the motion.

That is, when the wide scope readings of the quantifiers are true, the narrow scope readings are false.

The issue that we need to address is why the narrow scope reading of the quantifier is possible in cases like (6a) and (7a), but not in case like (8a) and (9a).

3 Logical Relations Between the Wide and Narrow Scope Readings

Intuitively, when we ask whether a sentence allows a wide or narrow reading, we mean whether the sentence is true in the conditions in which the wide or narrow reading is true. Wide scope interpretation is always available, unless it is excluded by some independent conditions. The narrow scope interpretation may but need not be true. For example, sentence (6a) allows the narrow scope reading of the quantifier since it is true in the conditions in which the narrow scope reading in (6b) is true. By contrast, sentence (8a) does not have the narrow scope reading, for it is false in the conditions in which the narrow scope reading in (8b) is true.

Consider now the sentence in (10a) [11] and the wide scope reading of the quantifier in (10b) and the narrow scope reading in (10c):

(10) a. Someone from New Jersey is likely to win the lottery.
b. **$\exists x$, from-New-Jersey** (x), **likely(*x* wins the lottery))** $\exists >$ *likely*
c. **likely($\exists x$, from-New-Jersey** (*x*), **(*x* wins the lottery))** *likely* $> \exists$

In a situation where one out of ten tickets is drawn in a lottery, three individuals from New York each bought one ticket, and one from New Jersey bought the remaining seven. Both the wide and narrow scope in (10b,c) readings are true. However, in a situation where seven individuals from New Jersey each bought one ticket, and one from New York bought the remaining three, the wide scope reading in (10b) is false, but the narrow scope reading in (10c) is true. Clearly it cannot be claimed that the narrow scope reading is derived exclusively from the wide scope reading, for it may be true when the wide scope reading is false.

We claim that a sentence allows for the narrow reading if logically it is materially implied $\rightarrow$ by the wide scope reading. This is not the same as claiming that the narrow scope reading is derived exclusively from the material implication from the wide scope reading. In our view, we first compute the wide and narrow scope readings of the sentence and then see if the wide scope reading being true guarantees the narrow scope reading being true. If it does, then the sentence allows for the narrow scope reading, i.e., the sentence is true under the conditions in which the narrow scope reading is true. This does not in any way exclude the case where the wide scope reading is false, but the narrow scope reading is true, for the material implication still holds.

Our view explains the difference between the sentences in (6a) and (7a) and those in (8a) and (9a) with respect to the narrow scope readings. On one hand, the wide scope readings of the sentences in (6a) and (7a) materially imply the narrow scope readings in (6b) and (7b) respectively; they thus allow for the narrow scope readings, i.e., the sentences in (6a) and (7a) are true in the conditions in which the narrow scope readings in (6b) and (7b) are true. On the other hand, the wide scope readings of the sentences in (8a) and (9a) do not materially imply the narrow scope readings in (8b) and (9b) respectively; hence, they do not allow for the narrow scope readings, i.e., the sentences in (8a) and (9a) are not true in the conditions in which the narrow scope readings in (8b) and (9b) are true.

Our analysis extends to the sentence in (11a) with the strong modal *is-guaranteed* [2]:

(11) a. Someone from New York is guaranteed to win the lottery.
b. It is guaranteed that someone from New York will win the lottery.

In the context in which the lottery tickets are exclusively sold to New York residents and each of them bought just one ticket, then the wide scope reading in (11a) is false, but the narrow scope reading in (11b) is true. However, much as in (10a), the wide scope reading materially implies the narrow scope reading, i.e., whenever the former is true, so is the latter. The sentence in (11a) thus allows for the narrow scope reading of the quantifier below the modal.

We believe that the same holds of non-monotonic quantifiers like *exactly n* where *n* is a natural number (we thank Tue Trinh for raising this point) and downward-entailing quantifiers like *at most n*:

(12) a. Exactly ten senators are likely to support the motion.
b. → It is likely that exactly ten senators support the motion.

(13) a. At most ten senators are likely to support the motion.
b. → It is likely that at most ten senators support the motion.

Here, too, the sentences in (12a) and (13a) allow for the narrow scope readings, for they are materially implied by the wide scope readings.

By contrast, when the truth of the wide scope reading does not materially imply that of the narrow scope reading, then the sentence does not allow for the narrow scope reading. The sentences in (14a) and (15a) with non-monotonic or downward-entailing quantifiers do not have the narrow scope readings in (14b) and (15b) respectively:

(14) a. Exactly ten senators are not likely to support the motion.
b. ↛ It is not likely that exactly ten senators support the motion.

(15) a. At most ten senators are not likely to support the motion.
b. ↛ It is not likely that at most ten senators support the motion.

The wide scope reading of sentence (14a) is true in a situation where exactly ten senators are likely to support the motion, exactly ten others are not, and the rest are undecided. But the narrow scope reading in (14b) is false in this situation. Hence, the sentence in (14a) does not allow for the narrow scope reading.

Some speakers find the wide and narrow scope readings of sentence (15a) with the downward-entailing quantifier *at most ten* a little difficult to process, but the difficulty can be removed by rephrasing them as in (16a) and (16b) respectively:

(16) a. The number of senators not likely to support the motion is at most ten.
b. ↛ It is not likely that the number of senators supporting the motion is at most ten.

In a situation where forty senators are likely to support and ten are not likely to, the sentence in (16a) is true, but that in (16b) is false. The sentence in (15a) thus does not allow for the narrow reading of the quantifier.

4 A-movement Scope Reconstruction

We consider here some examples of lack of A-movement scope reconstruction. We argue that they can be explained by the account according to which only the highest copy of an A-moved phrase is interpreted and by the idea that the narrow scope reading is possible if the wide scope reading materially implies it.

[6] takes the lack of the narrow scope reading in (17a) to be evidence that A-movement does not reconstruct (cf. Also [19]):

(17) a. Everyone seems <everyone> not to be here yet. $\forall > \neg$; $*\neg > \forall$
b. (it seems that) everyone isn't here yet. $\forall > \neg$; $\neg > \forall$

The argument is relatively straightforward. If the lower copy of the subject quantifier *everyone* is interpreted as in (17a), then the embedded clause is structurally the same as that in (17b). It is then predicted, incorrectly, that the quantifier may be in the scope of negation. However, if the copy is not interpreted, then whatever explains how negation may scope over the embedded subject will fail to apply, there being no interpreted copy in the embedded clause.[2] A-movement apparently does not reconstruct.

The sentence in (17a) has a reading in which the quantifier is in the scope of the modal *seem*, but has scope above negation. This reading is expected in our account. The wide scope reading in (18a) where the universal quantifier *everyone* takes scope above *seem* and negation materially implies the narrow scope reading in (18b) in which it is under the scope of *seem* but above negation:

(18) a. $\forall x$, **person** (x), **seem**$(\neg(x$ **is here yet**$))$
b. $\rightarrow$ **seem**$(\forall x$, **person** (x), $\neg(x$ **is here yet**$))$
c. $\nrightarrow$ **seem**$(\neg(\forall x$, **person** (x), x **is here yet**$))$

The sentence in (17a) is correctly predicted to have this reading. However, the reading in (18a) does not materially implies the reading in (18c) where the quantifier takes scope below both the modal *seem* and negation. This is hardly surprising, as the wide scope reading of the universal quantifier *everyone* over negation generally does not materially imply the narrow scope reading below negation, for any predicate **P**:

(19) a. $\forall x$ **person** (x), $\neg$**P**(x)
b. $\nrightarrow$ $\neg(\forall x$ **person** (x), **P**$(x))$

The sentence in (17a) therefore does not have the reading in which negation takes scope above the quantifier.

[20] argues that the example in (20a), in a situation in which five fair coins are flipped in a fair way, cannot be accurately paraphrased as in (20b):

(20) a. Every coin is 3% likely to land heads.
b. It is 3% likely that every coin will land heads.
c. Every coin is 3% likely [<every coin> to land heads]

If the sentence in (20a) is derived as in (20c) but the copy of the A-moved universal quantifier is not interpreted, then it follows that the quantifier is not in the scope of the matrix predicate. It is expected in our analysis that the sentence in (20a) where the

[2] [2] suggests that wide scope negation over a subject universal quantifier in the same clause is due to covert raising of negation. This operation may apply to (17b) with the subject quantifier is in the embedded clause, but not to (17a), for there is no interpreted copy of the raised subject in the embedded clause.

universal quantifier has wide cope over the matrix predicate does not have the narrow scope reading in (20b), since it does not materially imply it.

Consider now the contrast in (21) where the pronoun can only be understood to be a bound variable if the universal quantifier is A-moved to the matrix clause:

(21) a. Every professor$_i$ seems to his$_i$ students [<every professor$_i$ > to be smart]
b. *It seems to his$_i$ students that [that every professor$_i$ is smart]
c. **seem-to-his-students(**$\forall x$ **professor(**x**)(**x **is smart)**

The binding in (21a) is straightforward, as the highest copy of the A-moved subject is interpreted, thereby binding the pronoun that it c-commands. The lack of binding in (21b) may be explained syntactically, the quantifier not c-commanding the pronoun and quantifier scope being clause-bound. But our semantic account can account for it too.

The narrow scope reading of sentence (21a) is given in (21c). The pronoun is obviously not in the scope of the quantifier, and thus cannot be bound by it. The wide scope reading in (21a) with a bound pronoun therefore does not materially imply the narrow scope reading with an unbound pronoun. It is thus correctly predicted that the sentence does not have the narrow scope reading.

We now turn to A-movement reconstruction for binding condition C.

5 Late Merger and Scope Reconstruction

[21, 22] points out the binding contrast of the sort in (22) as evidence that adjuncts may attach to a noun phrase after it undergoes A-bar-movement, an option that is not available to complement PPs:

(22) a. [which pictures of John's room] does he like? *i=j; i≠j
b. [which pictures in John's room] does he like? i=j, i≠j

According to the copy theory of movement, the structures for the examples in (22a,b) are as in (23a,b) respectively:

(23) a. [which pictures [of John's room]] does he like <which pictures of John>?
b. [which pictures [in John's room]] does he like <which pictures>?

Since the complement PP *of John's room* in (23a) is part of the object before it moves, the proper name *John* in the copy left behind by the *wh*-phrase is bound, in violation of Binding Condition C. By contrast, the adjunct PP *in John's room* is not part of the object before it moves, but undergoes late merger, i.e., it is attached to it after movement, then the copy it leaves behind does not have the proper name that is bound as in (23b). The structure therefore does not violate Binding Condition C.

There is apparently no reason why late merger of adjuncts is not applicable to A-moved phrase. According to [29], the adjunct PP *with a hat* in (24a) is outside the scope of the modal *likely*:

(24) a. Some man with a hat is likely to arrive.
b. [some man [with a hat]] is likely [<some man> to arrive]

If the phrase *some man* is first merged [6] in the embedded clause and is raised to the matrix clause leaving behind a copy of itself as in (24b), and only then is the adjunct PP merged into it, it then follows that the PP is not in the scope of *likely*, for it is not part of the copy. In Ruys' analysis, technical details aside, the subject with the PP cannot reconstruct, since it is not identical to the copy.

The example in (25a), from [12], is another instance where late merger seems to block reconstruction, which in their judgement only has the reading in which the quantifier is outside the scope of the intensional predicate *look for*:

(25) a. I looked for a picture very intensely by this artist.
$\exists$ > *look for*; **look for* > $\exists$
b. I looked for [$_A$ a picture] very intensely [$_B$ a picture] [$_{PP}$ by this artist]

In [39], the indefinite *a picture* is first merged in its base-position and then moves to the right of the adverb *very intensely* (by overt QR [12]) and the adjunct PP *by this artist* is late-merged with *a picture*, as in (25b). The copy in A is realized at PF, and but the copy in B is interpreted. The extraposed phrase cannot reconstruct to A since it is not identical to it.

However, from the perspective of logical material implication, the sentence in (26a) materially implies that in (26b), contrary to the judgement reported by Ruys:

(26) a. Some man with a hat is likely to arrive. (=(24a))
b. → It is likely that some man with a hat is likely to arrive.

Similarly, the sentence in (27a) is like that in (25b) except that the copy *a picture* in A is not realized at PF, but that in B is both realized at PF and interpreted:

(27) a. I looked for very intensely a picture by this artist. (=(25b))
b. → I looked for one or other picture by this artist very intensely.

As the sentence in (27a) materially implies that in (27b), the narrow scope reading of the quantifier in (25a) ought to be possible as well.

In fact, there are some conceptual and empirical problems with Ruys' account for (24a) and (25a). Conceptually, late merger of adjuncts is optional [29]. The adjunct PP must undergo late merger in (22b) to circumvent a Binding Condition C violation. But there is no such motivation in (24a) and (25a). It should then in principle be possible for the adjunct PP in these cases to merge with the indefinite in its base-position. The subject undergoing raising in (26a) and the extraposed phrase in (27a) would leave behind identical copies. It should be possible to interpret these copies giving rise to the narrow scope readings.

Empirically, the wide scope reading of the indefinite in (27a) is logically equivalent to the sentence in (28), which materially implies the sentence in (27b):

(28) There is a picture by this artist that I looked for very intensely.

The sentence in (27a) therefore materially implies the sentence in (27b). There is then no reason why it does not allow for the narrow scope reading of the indefinite.

Moreover, additional evidence for an indefinite structurally higher than an intensional predicate to have scope over it comes from the example in (29a):

(29) a. I looked for many times a picture by this artist.
b. I looked for a picture by this artist many times.

In (29a), the frequency adverb *many times* semantically has scope over the intensional predicate *look for*, i.e., many times of looking for; hence, it is structurally higher than the predicate. The extraposed indefinite *a picture by this artist* is structurally higher than the adverb, and is thus higher than the predicate. Yet, semantically it may be in the scope of both, just as it does when it is syntactically in its base-position in (29b). The sentence in (29a) can mean that I looked for one or other picture by this artist, much like that in (29b). The narrow scope reading of the indefinite in (29a) is hardly surprising, for it is materially implied by the wide scope reading. If it is true that there is a picture by this artist that I looked for many times, then it is also true that I looked for one or other picture by this artist many times.

The wide scope reading of the indefinite in (25a) and (27a) recalls the de re/de dicto readings of an indefinite with respect to the intensional predicate. In fact, there is reason to believe that the extraposed indefinite in these cases may be in the scope of *look for*. [2] credits the examples in (30) to Gennaro Chierchia:

(30) a. Even though there are no unicorns, yet a unicorn seems to be approaching.
b. Even though unicorns are fictional entities, yet a unicorn seems to be approaching.

These sentences are not self-contradicting, for the indefinite *a unicorn* is understood to be in the scope of the modal *seem*. They apparently show that an indefinite syntactically outside the scope of a modal may be read de dicto in the scope of the modal. If so, then the indefinite in (27a) and (29a), too, despite its being structurally outside the scope of the modal, much like the indefinite in (30), may have a de dicto reading in the scope of the intensional predicate *look for* as well.

6 Binding and Interpretation of Formal Features

Certain facts about binding seem to present a problem for our conclusion that only the highest copy of an A-moved phrase is interpreted.

For our immediate concerns here, we consider the basic cases in (31)–(32), familiar from the study of binding [3, 4]:

(31) a. John believed [Mary$_i$ to have criticized herself$_i$]
b. *John$_i$ believed [Mary to have criticized himself$_i$]

(32) a. *John believed [Mary$_i$ to have criticized her$_i$]
b. John$_i$ believed [Mary to have criticized him$_i$]

Binding Condition A requires that an anaphor be bound in the embedded clause, whence the contrast in (31). Binding Condition B requires that a pronoun be free in the embedded clause, whence the contrast in (32).[3]

In this light, it comes as a surprise that in (33a) the anaphor may be bound by the subject in the matrix clause:

(33) a. It seemed [that John$_i$ has criticized himself$_i$]
b. John$_i$ seemed [to have criticized himself$_i$] (cf. (31b))
c. John$_i$ seemed [<John$_i$> to have criticized himself$_i$]

The subject in the matrix clause in (33b) behaves as if it is in the embedded clause, for it may bind the anaphor. The binding can be accounted for, if the matrix subject leaves a copy of itself in the embedded clause when it moves to the matrix clause, as in (33c). If the lower copy is interpreted, then it can bind the anaphor. But this is impossible on the view according to which only the highest copy of an A-moved phrase is interpreted.

The same problem arises in (34). The matrix subject in (34b), too, behaves as if it is in the embedded clause, illicitly binding the pronoun violating Binding Condition B:

(34) a. *It seemed [that John$_i$ has criticized him$_i$]
b. *John$_i$ seemed [to have criticized him$_i$] (cf. (32b))
c. *John$_i$ seemed [<John$_i$> to have criticized him$_i$]

Interpreting the lower copy if the A-moved subject explains why the pronoun is illicitly bound in the embedded clause. Again, this is impossible if only the highest copy of an A-moved phrase is interpreted.

Facts concerning binding by a quantifier are essentially the same. The quantifier subject A-moved to the matrix clause may bind an anaphor and may not bind a pronoun in the embedded clause:

[3] For the definition of the local domain for binding, see [4]. The domains in which Binding Conditions A and B are satisfied do not always coincide [14]. For example, both an anaphor and a pronoun in a nominal phrase may be bound by an antecedent in the same clause.

(35) a. Some man$_i$ seemed [to have criticized himself$_i$]
b. Some man$_i$ seemed [<some man$_i$> to have criticized himself$_i$]

(36) a. *Some man$_i$ seemed [to have criticized him$_i$]
b. *Some man$_i$ seemed [<some man$_i$> to have criticized him$_i$]

It looks as if we must interpret the copy in the embedded clause in order to account for the binding facts.

The examples in (37a) and (38a) present an especially serious problem for our conclusion in the last section that only the highest copy of an A-moved phrase is interpreted:

(37) a. Some senator$_i$ is not likely to criticize himself$_i$
b. Some senator$_i$ is not likely [<some senator$_i$ > criticizes himself$_i$]

(38) a. *Some senator$_i$ is not likely to criticize him$_i$
b. *Some senator$_i$ is not likely [<some senator$_i$ > criticizes him$_i$]

On one hand, if the lower copy is not interpreted, then it becomes a mystery as to how the anaphor in (37a) is bound and how the pronoun in (38a) is not free in the embedded clause. On the other hand, if we interpret the lower copies to explain the binding facts, then it would lead to incorrect interpretations of the scope of negation, as shown in the last section.

To resolve this dilemma, we would like to suggest that formal features be divided into lexical features (l-features) and morphosyntactic features (ms-features). As our interests here concern binding and scope of quantifiers, we will limit ourselves to formal features of nominals.

A nominal expression essentially has two kinds of formal features: l-features and ms-features. The former comprises property-denoting features (*student*, *teacher*, *picture* etc.) and semantic features distinguishing *the*, *a*, *some*, *that*, etc. The latter comprises categorial, person, number and gender features. Thus, two nominals may differ with respect to the l-features but are indistinguishable with respect to the ms-features, or they may be the same with respect to l-features but differ from each other with respect to the ms-features. Thus, in (39a) *student* and *teacher* differ with respect to l-features, but are the same with respect to ms-features:

(39) a. student, teacher, picture, …
b. student, students; teacher, teachers; picture, pictures, …

In (39b), *student* and *students* differ with respect to ms-features, but are the same with respect to l-features.

For binding principles, what is relevant are the ms-features, not the l-features. The difference in l-features has no relevance to binding principles. Thus, binding is just the same whether the antecedent is *boy* or *male teacher* or is *girls*, *female teachers* even though members of each pair differ from each other with respect to the l-features:

(40) a. {Some, the} {boy$_i$, male teacher$_i$} criticized {himself$_i$, *themselves$_i$, *herself$_i$}

b. {*That, those} {girls$_i$, female teachers$_i$} criticized {*himself$_i$, *herself$_i$, themselves$_i$}

(41) a. *{Some, the} {boy$_i$, male teacher$_i$} criticized him$_i$.
b. *Those {girls$_i$, female teachers$_i$} criticized them$_i$.

Accordingly, the notion of binding in Binding Theory [3, 4] should be revised as in (42):

(42) A binds B iff
a. A c-commands B, and
b. their person, number and gender ms-features have the same values.

An expression is free if it is not bound. Binding has no requirement regarding l-features.

We propose a dynamic theory of interpretation according to which formal features are interpreted according to their relevance to particular grammatical principles. As ms-features are relevant to binding (see (42)) but l-features are not, the former but not the latter are interpreted where Binding principles apply. L-features are generally interpreted after all syntactic operations.

In cases without movement, the selective interpretation of formal features makes little difference. Its significance shows up in cases with movement:

(43) a. {John$_i$, some man$_i$} seemed [to have criticized himself$_i$]
b. {John$_i$, some man$_i$} seemed [{<John$_i$>/<some man$_i$>} to have criticized himself$_i$]

(44) a. *{Those, at most five} {girls$_i$, female teachers$_i$} seemed [to have criticized them$_i$]
b. *{Those, at most five} {girls$_i$, female teachers$_i$} seemed [{<those>/<at most five>} {<girls$_i$>/<female teachers$_i$>} to have criticized them$_i$]

The binding facts in these examples can be accounted for by interpreting the ms-features in the copies left behind by the A-moved subject, since these features are relevant to binding (see (42)).

Along these lines, we can resolve the requirements by Binding principles and lack of A-movement reconstruction. In (45a) and (46a) the interpretations of the ms-features of the copies left behind by the A-moved subject explain why the anaphor is bound and the pronoun is free in the embedded clause; they have the same person, number and gender ms-features as those of their antecedents:

(45) a. Some senator$_i$ is not likely to criticize himself$_i$
b. Some senator$_i$ is not likely [<some senator$_i$ > to criticize himself$_i$]

(46) a. *Some senator$_i$ is not likely to criticize him$_i$
b. *Some senator$_i$ is not likely [<some senator$_i$ > to criticize him$_i$]

As the l-features of the A-moved subject *some senator* are not relevant to binding, they are not interpreted in the embedded clause. As a result, it does not fall in the scope of negation, as desired. The l-features in the highest copy are interpreted, yielding the readings in which the A-moved subject has scope above both negation and the matrix predicate *likely*.

The examples in (47) and (48) [11] apparently suggest that A-movement may reconstruct (the copies in angle-brackets are added for our purposes here):[4]

(47) a. A student of David$_i$'s seems to him$_i$ [<a student of David$_i$'s> to be at the party] $\exists>$*seem*; *$seem>\exists$
b. A student of his$_i$ seems to David$_i$ [<a student of his$_i$> to be at the party] $\exists>$*seem*; $seem>\exists$

The lack of the narrow scope reading in (47a) can be accounted for if the copy in the embedded clause the A-moved subject leaves behind is interpreted, in violation of Binding Condition C. The example in (47b) is fine, regardless of whether the copy in the embedded clause is interpreted, as the pronoun is free in the DP containing it (see note 3).

We would like to suggest that Binding Conditions be satisfied at the point of derivation in which the domains relevant to them are constructed. For instance, for the example in (33b), Binding Condition A is satisfied at the point where the embedded clause is built as in (33c). Likewise, Binding Condition B in the example in (34b) is not satisfied at the point when the embedded clause is built in (34c). The same process applies in the structures in (35b), (36b), (37b) and (38b).

Differing from other Binding Conditions A and B, Binding Condition C requires that a proper name be free, without regard to the locality of the antecedent. Along these lines, Binding Condition C is violated in (47a) at the point in the derivation in which the embedded clause is merged with the dative *to him*, before the embedded subject moves the the matrix clause.

The contrast in (48) [11] too seems to indicate that A-movement may reconstruct (the copies in angle-brackets are added for our purposes here):

(48) a. ??Someone from his$_i$ class shouted to every professor$_i$ [PRO$_i$ to be careful]
b. Someone from his$_i$ class seems to every professor$_i$ [<someone from his$_i$ class> to be careful]

[4] We treat proper names here as property-denoting, e.g., *David* denotes a set of properties that David has. (cf. [17]). In some Romance languages, proper names co-occur with determiners much like common nouns. English allows it to some limited extent, e.g., *the David that I know*.

The bound reading of the pronoun in (48b) can be accounted for by interpreting the copy the A-moved subject leaves behind in the embedded clause. The lack of the bound variable reading of the pronoun in the example in (48a) with a matrix control verb is precisely because there is no copy in the embedded clause. It is not possible to derive to the bound reading in (48b) by QR of the quantifier, for that would incur the same Weak Crossover [24] violation as that in (48a) [18, 25, 28].

We would like to suggest that as variable binding, like Binding Condition C, bears on the interpretation of ms-features and is not subject to locality, apart from structural c-command, a pronominal variable can be bound by an antecedent c-commanding it at any point in the derivation. Along these lines, the pronoun in the embedded subject can be bound by the universal quantifier at the point where the embedded clause is merged with the dative *to every professor* in the matrix clause. The binding relationship can be formally indicated by the familiar coindexing. It is this binding relationship that allows the pronoun to be bound as a variable by the universal quantifier.

We note here that the matrix subject with the bound variable reading of the pronoun in (47b) appears to take scope above *seem*. This reading cannot be explained if the lower copy of the A-moved subject is interpreted, for the subject would necessarily be in the scope of *seem*.

7 Conclusions

In our investigation of A-movement of quantifiers undertaken here, we show that syntactic scope reconstruction, via the copy theory of movement, may lead to some incorrect results.

We claim that relative scope of quantifiers, modals and negation is derived from the logical relations among them, specifically, from the logical material implications. As logical relations are independent of A-movement, our analysis relying on these and requiring no additional assumptions can be said to bear no theoretical cost specific to A-movement.

We argue that the same logical relations lead us to the consequence that late merger of adjuncts has no effect on quantifier scope in extraposition. We show that a quantifier with a late-merged adjunct structurally higher than an intensional predicate may nevertheless have scope below it. If our account is correct, then scope interpretation is essentially semantic.

Our suggestion that formal features be interpreted according to their relevance to the grammatical principles at the point they apply resolves some conflicting demands by binding and scope. Formal features are not interpreted all at the same time as is commonly taken for granted, but may be selectively interpreted. On one hand, ms-features bearing on syntactic principles are interpreted derivationally, i.e., as the syntactic structures are derived by successive merge [6]. The features are interpreted when the structural constraints they are subject to become relevant, possibly phase-wise [7, 8], to reduce the amount of checking. On the other hand, l-features in the highest copy of an A-moved phrase are interpreted when all movement has taken place, for it may cross scope-bearing expressions.

Our analysis thus enlarges the body of evidence showing lack of reconstruction in A-movement.

Acknowledgement. We gratefully acknowledge the generous funding of the present work under General Research Fund Grant #11606418 by Research Grants Council of the government of the Hong Kong Special Administrative Region. We would like to thank Xue Bo for clarifying several logical issues as well as two anonymous reviewers for helpful comments on an earlier version of the paper.

References

1. Barss, A.: Chains and Anaphoric Dependence. PhD dissertation. MIT (1986)
2. Boeckx, C.: Scope reconstruction and A-movement. Nat. Lang. Linguist. Theory **19**, 503–548 (2001)
3. Chomsky, N.: Lectures on Government and Binding. Foris, Dordrecht (1981)
4. Chomsky, N.: Knowledge of Language. Praeger, New York (1986)
5. Chomsky, N.: A minimalist program for linguistic theory. In: Hale, K., Keyser, S. (eds.) The View from Building 20: Essays in Linguistics in Honor of Sylvain Bromberger, pp. 1–52. MIT Press, Cambridge (1993)
6. Chomsky, N.: The Minimalist Program. MIT Press, Cambridge (1995)
7. Chomsky, N.: Derivation by phase. In: MIT Working Papers in Linguistics, vol. 19. MIT, Cambridge (1999)
8. Chomsky, N.: On phases. In: Freidin, R., Otero, C.P., Zibizaretta, M.L. (eds.) Foundational Issues in Linguistic Theory. Essays in Honor of Jean-Roger Vergnaud, pp. 291–321. MIT Press, Cambridge (2008)
9. Cresti, D.: Extraction and reconstruction. Nat. Lang. Seman. **3**, 283–341 (1995)
10. Fox, D.: Reconstruction, variable-binding and the interpretation of chains. Linguist. Inq. **30**, 157–196 (1999)
11. Fox, D.: Economy and Semantic Interpretation. MIT Press, Cambridge (2000)
12. Fox, D., Nissenbaum, J.: Condition a and scope reconstruction. Linguist. Inq. **35**, 475–485 (2004)
13. Heycock, C.: Asymmetries in reconstruction. Linguistic Inq. **26**, 547–570 (1995)
14. Huang, C.-T.J.: On the distribution and reference of empty pronouns. Linguist. Inq. **15**, 531–574 (1984)
15. Huang, C.-T.J.: Reconstruction and the structure of VP: some theoretical consequences. Linguist. Inq. **24**, 103–138 (1993)
16. Iatridou, S., Sichel, I.: Negative DPs, A-movement and scope diminishment. Linguist. Inquiry **42**, 595–629 (2011)
17. Keenan, E.: Eliminating the universe a study in ontological perfection. In: Flickinger, D., et al. (eds.) Proceedings of the first West Coast Conference on Formal Linguistics, pp. 71–81. CSLI publications, Stanford (1982)
18. Koopman, H., Sportiche, D.: Variables and the Bijection principle. Linguist. Rev. **2**, 139–160 (1982)
19. Lasnik, H.: Chains of arguments. In: Epstein, S., Hornstein, N. (eds.) Working Minimalism, pp. 189–215. MIT Press, Cambridge (1999)
20. Lasnik, H.: Minimalist Investigations in Linguistic Theory. Routledge, London/New York (2003)
21. Lebeaux, D.: Language acquisition and the form of grammar. PhD dissertation, University of Massachusetts, Amherst (1988)
22. Lebeaux, D.: Relative clauses, licensing and the nature of derivations. In: Rothstein, S. (ed.) Perspectives on Phrase Structure: Heads and Licensing, pp. 209–239. Academic Press, San Diego (1991)

23. Lechner, W.: Two kinds of reconstruction. Stud. Linguist. **52**, 276–310 (1998)
24. Postal, P.: Cross-Over Phenomena. Holt, Rinehart and Winston, New York (1971)
25. Postal, P.: Remarks on weak crossover effects. Linguist. Inq. **24**, 539–556–539 (1993)
26. van Riemsdijk, H., Williams, E.: NP Struct. Linguist. Rev. **1**, 171–217 (1981)
27. Romero, M.: Problems for a semantic account of scope reconstruction. In: Katz, G., Kim, S., Winhart, H. (eds), Reconstruction: Proceedings of the 1997 Tübingen Workshop Arbeitspapiere des Sonderforschungsbereichs 340, Berich Nr. 127, Universität Stuttgart and Universität Tübingen, pp. 119–146 (1998)
28. Ruys, E.G.: Weak crossover as a scope phenomenon. Linguist. Inq. **31**, 513–539 (2000)
29. Ruys, E.G.: A minimalist condition on semantic reconstruction. Linguist. Inq. **46**, 453–488 (2015)
30. Sauerland, U.: Scope reconstruction without reconstruction. In: Shahin, K., Blake, S., Kim, E. (eds.) Proceedings of the Seventeenth West Coast Conference on Formal Linguistics, pp. 582–596. CSLI Publications, Stanford (1999)
31. Sportiche, D.: Reconstruction, binding and scope. In: Everaert, M., van Riemsdijk, H. (eds.) The Blackwell Companion to Syntax, vol. IV, pp. 35–93. Blackwell, Oxford (2006)

Acts of Commanding and Promising in Dynamified Common Sense Term-Sequence-Deontic-Alethic Logic

Katsuhiko Sano(✉) and Tomoyuki Yamada

Faculty of Humanities and Human Sciences, Hokkaido University, Nishi 7 Chome, Kita 10 Jo, Kita-ku, Sapporo, Hokkaido 060-0810, Japan
v-sano@let.hokudai.ac.jp, yamada@hokkaido.email.ne.jp

Abstract. The language of propositional modal logic has been shown highly useful in developing logics of various specific speech acts. But it is also clear that we need more expressive language if we wish to say, for example, that, for any x and y, whenever x promises y to see to it that y is safe, it becomes obligatory for x to see to it that y is safe with respect to y in the name of x. We develop a dynamic modal predicate logic, **DCTSDAL**$^=$ (Dynamified Commonsense Term-Sequence-Deontic-Alethic Logic with equality), in which we can assert exactly the kind of things like this by extending a version of term-sequence-modal logic introduced in [7] with dynamic modalities that stand for acts of commanding and promising. We develop a semantics which incorporates the commonsense treatment of free variables introduced in [1] and combine it with the varying domain assumption as is done in [9]. The main contribution of this paper is the relative semantic completeness of **DCTSDAL**$^=$ with respect to its static base logic **CTSDAL**$^=$ via recursion axioms.

Keywords: Speech act · Command · Promise · Dynamic modal predicate logic · Term-sequence-modal logic · Relative completeness

1 Introduction

The language of propositional modal logic has been shown highly useful in developing logics of various specific speech acts. But it is also clear that we need more expressive language if we wish to say, for example, that if you promise a person to keep her safe, you are committed to keep her safe. Its natural formalization might be something like the following:

$$\forall x. \forall y. [\mathsf{Prom}_{(x,y)}\mathrm{Safe}(y)]\mathsf{O}_{(x,y,x)}\mathrm{Safe}(y)$$

where $[\mathsf{Prom}_{(x,y)}\psi]\varphi$ means that whenever x promises y that x will see to it that ψ, φ holds in the resulting situation and $\mathsf{O}_{(x,y,z)}\varphi$ means that it is obligatory for x (the agent who owes the obligation) with respect to y (the agent to whom the obligation is owed) in the name of z (the agent who creates the obligation) to see to it that φ. In this paper, we will develop a logic in which we can assert exactly the kind of things like this.

D. Deng et al. (Eds.): TLLM 2022, LNCS 13524, pp. 116–135, 2023.
https://doi.org/10.1007/978-3-031-25894-7_6

The reader may wonder why we need to have three indices for our deontic operator $\mathsf{O}_{(x,y,z)}$ above. This is because we analyze not only acts of promising but also acts of commanding. For example, if you command a person to keep herself safe, she is obligated to keep herself safe. Similarly as above, this may be naturally formalized as follows:

$$\forall x. \forall y. [\mathsf{Com}_{(x,y)}\mathsf{Safe}(y)]\mathsf{O}_{(y,x,x)}\mathsf{Safe}(y),$$

where $[\mathsf{Com}_{(x,y)}\psi]\varphi$ means that whenever x commands y to see to it that ψ, φ holds in the resulting situation. In the case of an act of commanding, the agent who creates the obligation is the commander and is identical with the agent to whom the obligation is owed (sometimes called "obligee"), while the agent who owes the obligation (sometimes called "obligor") is the commandee. In the case of an act of promising, however, the agent who creates the obligation is the agent who makes the promise (the promiser) and is identical with the agent who owes the obligation while the agent to whom the obligation is owed is the agent to whom the promise is given (the promisee). If we only includes the roles of the creator of the obligation and the agent who owes the obligation, the role of the agent to whom the promise is given would be ignored, which does not seem right.

These two speech acts were already studied in [13] in the propositional setting, but this previous study cannot explicitly capture an dependency of y in $\mathsf{Safe}(y)$ on, say, the act $[\mathsf{Com}_{(x,y)}\mathsf{Safe}(y)]$ of commanding or the deontic operator $\mathsf{O}_{(y,x,x)}$ as it does not have the distinction between terms and predicates. Note that all occurrences of the varibles x and y in the formula $[\mathsf{Com}_{(x,y)}\mathsf{Safe}(y)]\mathsf{O}_{(y,x,x)}\mathsf{Safe}(y)$, including the occurrences of x and y as the indices of $\mathsf{Com}_{(x,y)}$ and $\mathsf{O}_{(y,x,x)}$, are bound by the universal quantifiers $\forall x.$ and $\forall y.$ in the formula $\forall x. \forall y. [\mathsf{Com}_{(x,y)}\mathsf{Safe}(y)]\mathsf{O}_{(y,x,x)}\mathsf{Safe}(y)$ respectively.

In formalizing the above examples, we also incorporate the intuitive idea of the contingency of existence of agents like us. I do exist in the actual world, say w, but I am not a necessary being and so there is a possible world v such that v is accessible from w but I do not exist in v. I actually do not have a sister, but it is not necessary that I do not have a sister; there is a possible world u such that u is accessible from w and someone who does not exist in w is my sister in u. One straightforward way of representing such contingency of existence of individual entities is to assume that the domains of possible worlds vary in any ways unless they are not empty. We refer to this assumption as varying domain assumption (VDA, for short).

We introduce quantification by recasting the dynamic logic of acts of commanding and promising developed in [13] in a dynamic extension of the language of Term-sequence-modal Logics (**TSMLs**) developed in [7].[1] **TSMLs** are generalizations of Term-Modal Logics (**TMLs**) developed by [4, 10] in the sense that **TMLs** are instances of **TSML**. [2] For, in **TMLs**, modalities are allowed to be indexed by a term; for example, the statement that every Christian believes in the existence of God can be expressed by

[1] The second author has developed a **PAL**-style dynamic logic that characterizes effects of acts of requesting and asserting along with those of acts of commanding and promising in [15] in the propositional setting, but in order to keep things simple, we will ignore acts of requesting and asserting in this paper. This enables us to concentrate on alethic and deontic readings of term-sequence modal operators.

[2] Modal predicate logics are also instances of **TSML**.

the following formulas [4, p. 134]):

$$\forall x.(\text{Christian}(x) \to [x]\exists y.\text{God}(y)),$$

where $[x]\varphi$ is read as "x believes that φ". **TSMLs** extend **TMLs** by allowing modal operators to be indexed by a finite sequence of terms. Thus, the moral judgement that parents are obligated to keep their young children safe can be expressed by the following:

$$\forall x.\forall y.((\text{Parent}(x,y) \land \text{Young}(y)) \to \mathsf{O}_{(x,y,x)}\text{Safe}(y)).$$

And finally, by defining model updating operations, we may introduce dynamic operators standing for types of acts of commanding and promising. Then it becomes possible to express, for example, the judgement that if you promise a person to keep her safe, you are committed to keep her safe by the following formula mentioned at the beginning of this paper:

$$\forall x.\forall y.[\mathsf{Prom}_{(x,y)}\text{Safe}(y)]\mathsf{O}_{(x,y,x)}\text{Safe}(y).$$

Another policy we will incorporate in the logic to be developed in this paper is that of common sense treatment of free variables (CTFV, for short), proposed in [1] and further developed in [9]. It requires free variables in the formula to be interpreted by objects existing in the local domain at the current world (see [1, p.121]).

One consequence of the combination of CTFV and VDA is that K-axiom $\Box(\varphi \to \psi) \to (\Box\varphi \to \Box\psi)$ is not valid. But, if $\mathsf{FV}(\varphi) \subseteq \mathsf{FV}(\psi)$ holds then K-axiom above is valid, where $\mathsf{FV}(\varphi)$ is the set of free variables occurring in φ. As a result, we need to take care of free variables when we apply replacement of equivalent formulas. That is, when φ and ψ are equivalent and $\mathsf{FV}(\varphi) = \mathsf{FV}(\psi)$, a formula χ and a resulting formula replacing some occurrences of φ in χ with ψ are equivalent. When we turn our eyes on dynamic epistemic logic, replacement of equivalent formulas plays an important role in applying the (inside-out) recursion strategy of reducing the semantic completeness of, say, public announcement logic to the semantic completeness of the underlying epistemic logic (cf. [12]). In this sense, it is a non-trivial question if we can apply a recursion strategy also in our setting. We answer this question positively.

The outline of this paper is as follows. Section 2 presents the syntax which has deontic and alethic modal operators as well as the equality symbol. Note that the equality symbol is needed in order to state recursion axioms for its dynamic extension.[3] Then we move on to the semantics which incorporates VDA and CTFV. We also provide our Hilbert system H(**CTSDAL**$^=$), which is sound for our semantics (Theorem 1). In Sect. 3, we show how **CTSDAL**$^=$ can be extended into a dynamic logic, **DCTSDAL**$^=$, by introducing dynamic modalities that stand for types of acts of commanding and promising. We also specify which acts of commanding and promising generate obligations. In particular, when the content of acts of commanding and promising is written in terms of a first-order formula, they always generate obligations (Theorem 2). This is a similar result to that for the notion of successful formulas in the public announcement logic (cf. [6,11]). In Sect. 4, we establish that **DCTSDAL**$^=$ is sound (Theorem 3) and semantically complete relatively to **CTSDAL**$^=$ in the sense that **DCTSDAL**$^=$ is semantically complete if **CTSDAL**$^=$ is semantically complete (Theorem 4). Section 5 concludes the paper.

---

[3] Recursion axioms are also known as "reduction axioms" in literature.

2 Syntax, Semantics and Hilbert System of CTSDAL$^=$

The syntax $\mathcal{L}$ for **CTSDAL$^=$** consists of the following vocabulary: (i) a countably infinite set $\mathsf{Var} = \{x, y, \ldots\}$ of *variables*, (ii) a countably infinite set $\mathsf{Pred} = \{P, Q, \ldots\}$ of *predicate symbols*, each of which has a fixed finite arity, and (iii) the equality symbol $=$, (iv) a set of logical symbols whose members are $\neg$, $\rightarrow$, $\forall$, $\Box$ and $\mathsf{O}_{(x,y,z)}$ where $x, y, z \in \mathsf{Var}$. Then, the set $\mathsf{Form}_{\mathcal{L}}$ of all formulas are inductively defined as follows:

$$\mathsf{Form}_{\mathcal{L}} \ni \varphi ::= P(x_1, \ldots, x_n) \mid x = y \mid \neg\varphi \mid (\varphi \rightarrow \varphi) \mid \forall x.\, \varphi \mid \Box\varphi \mid \mathsf{O}_{(x,y,z)}\varphi,$$

where $P \in \mathsf{Pred}$ and all of $x_1, \ldots, x_n, x, y, z$ are variables in Var. We read $\Box\varphi$ as "it is necessary that φ" and $\mathsf{O}_{(x,y,z)}\varphi$ as "it is obligatory for x with respect to y in the name of z to see to it that φ." We define other Boolean connectives such as conjunction $\wedge$, disjunction $\vee$ and logical equivalence $\leftrightarrow$ as usual and we abbreviate $\neg(x = y)$ as $x \neq y$. We also use the defined dual $\mathsf{P}_{(x,y,z)}\varphi := \neg\mathsf{O}_{(x,y,z)}\neg\varphi$, which is read as "$x$ is permitted with respect to y in the name of z to see to it that φ". Let us also introduce the following abbreviation, which is useful to state recursion axioms later: Given a finite set $X = \{x_1, \ldots, x_n\}$ of variables, we define

$$\top_X := P(x_1, \ldots, x_n) \rightarrow P(x_1, \ldots, x_n)$$

Thus, $\top_X$ is a tautology written in terms of all variables in X.[4] We define the complexity $\mathsf{c}(\varphi)$ of a formula φ as the total number of occurrences of logical connectives in φ.

Definition 1. *Given a formula φ, we define the set $\mathsf{FV}(\varphi)$ of all free variables in φ inductively as follows:*

$$\begin{aligned}
\mathsf{FV}(P(x_1, \ldots, x_n)) &:= \{x_1, \ldots, x_n\}, \\
\mathsf{FV}(x = y) &:= \{x, y\}, \\
\mathsf{FV}(\neg\varphi) &:= \mathsf{FV}(\varphi), \\
\mathsf{FV}(\varphi \rightarrow \psi) &:= \mathsf{FV}(\varphi) \cup \mathsf{FV}(\psi), \\
\mathsf{FV}(\forall x.\, \varphi) &:= \mathsf{FV}(\varphi) \setminus \{x\}, \\
\mathsf{FV}(\Box\varphi) &:= \mathsf{FV}(\varphi), \\
\mathsf{FV}(\mathsf{O}_{(x,y,z)}\varphi) &:= \{x, y, z\} \cup \mathsf{FV}(\varphi).
\end{aligned}$$

We use $\varphi[y/x]$ as the result of substituting all free occurrences of x in φ uniformly with y, where we avoid the variable clash and rename the bound variable to a fresh variable.

We move on to our semantics. A *model* M is a tuple

$$M := (W, D, R_{\Box}, R_{\mathsf{O}}, V)$$

where W is a non-empty set of *possible worlds*; D is a family $(D(w))_{w \in W}$ of non-empty set of objects; $R_{\Box}$ is a binary relation on W; R_{O} is a family of binary relations $R_{\mathsf{O}}(d_1, d_2, d_2) \subseteq W \times W$ indexed by $(d_1, d_2, d_3) \in (\bigcup_{w \in W} D(w))^3$ such that $R_{\mathsf{O}}(d_1, d_2, d_3) \subseteq R_{\Box}$; V assigns each predicate symbol P with $V(P, w) \subseteq D(w)^n$ where the arity of P is n.

As in the following definition, our variable assignment is *rigid*, i.e., it does not depend on a given world.

[4] We may also define $\top_X := (x_1 = x_1) \wedge \cdots \wedge (x_n = x_n)$.

Definition 2. *Let* $M = (W, D, R_{\Box}, R_{\bigcirc}, V)$ *be a model. A* variable assignment *is a function* $\alpha : \mathsf{Var} \to \bigcup_{w \in W} D(w)$. *We define* $\alpha(x|d)$ *as the same variable assignment as* α *except sending* x *to* $d \in \bigcup_{w \in W} D(w)$. *Given a variable assignment* α, *a formula* φ *is said to be an* α_w-formula *if, for all* $x \in \mathsf{FV}(\varphi)$, $\alpha(x) \in D(w)$ *holds, i.e.,* $\alpha[\mathsf{FV}(\varphi)] \subseteq D(w)$, *where* $\alpha[Y]$ *is the direct image of* $Y \subseteq \mathsf{Var}$ *under the function* α.

Definition 3. *Let* $M = (W, D, R_{\Box}, R_{\bigcirc}, V)$ *be a model and* α *a variable assignment. We define the* satisfaction relation $M, \alpha, w \models \varphi$ *between* M, α, w *and* α_w*-formula* φ *inductively as follows:*

$$
\begin{aligned}
M,\alpha,w \models P(x_1,\ldots,x_n) &\text{ iff } (\alpha(x_1),\ldots,\alpha(x_n)) \in V(P,w),\\
M,\alpha,w \models x = y &\text{ iff } \alpha(x) = \alpha(y),\\
M,\alpha,w \models \neg\varphi &\text{ iff } M,\alpha,w \not\models \varphi,\\
M,\alpha,w \models \varphi \to \psi &\text{ iff } M,\alpha,w \models \varphi \textit{ implies } M,\alpha,w \models \psi,\\
M,\alpha,w \models \forall x.\varphi &\text{ iff } \textit{for all } d \in D(w), M,\alpha(x|d),w \models \varphi,\\
M,\alpha,w \models \Box\varphi &\text{ iff } \textit{for all } v \in W((w,v) \in R_{\Box} \textit{ and } \alpha[\mathsf{FV}(\varphi)] \subseteq D(v)\\
&\qquad \textit{imply } M,\alpha,v \models \varphi),\\
M,\alpha,w \models \mathsf{O}_{(x,y,z)}\varphi &\text{ iff } \textit{for all } v \in W((w,v) \in R_{\bigcirc}(\alpha(x),\alpha(y),\alpha(z)) \textit{ and } \alpha[\mathsf{FV}(\varphi)] \subseteq D(v)\\
&\qquad \textit{imply } M,\alpha,v \models \varphi).
\end{aligned}
$$

We say that φ *is* valid *if, for all models* $M = (W, D, R_{\Box}, R_{\bigcirc}, V)$, *worlds* $w \in W$ *and variable assignments* α *such that* $\alpha[\mathsf{FV}(\varphi)] \subseteq D(w)$, $M, \alpha, w \models \varphi$ *holds.*

It is noted that $\mathsf{O}_{(x,y,z)}\varphi$ is an α_w-formula in the satisfaction clause for $\mathsf{O}_{(x,y,z)}\varphi$ and so all of $\alpha(x)$, $\alpha(y)$ and $\alpha(z)$ are elements of $D(w)$. For a formula of the form $\mathsf{P}_{(x,y,z)}\varphi$, we can obtain the following satisfaction clause:

$$
\begin{aligned}
M,\alpha,w \models \mathsf{P}_{(x,y,z)}\varphi \text{ iff } &\text{for some } v \in W((w,v) \in R_{\bigcirc}(\alpha(x),\alpha(y),\alpha(z)) \text{ and } \alpha[\mathsf{FV}(\varphi)] \subseteq D(v)\\
&\text{and } M,\alpha,v \models \varphi).
\end{aligned}
$$

Seligman [9] observed that the Barcan formula $\forall x.\Box\varphi \to \Box\forall x.\varphi$ is not valid but the converse Barcan formula $\Box\forall x.\varphi \to \forall x.\Box\varphi$ is valid. While K-axiom $\Box(\varphi \to \psi) \to (\Box\varphi \to \Box\psi)$ is not valid, it is valid if $\mathsf{FV}(\varphi) \subseteq \mathsf{FV}(\psi)$ holds. It is noted that $\varphi \to \psi$ is called an *involvement* when $\mathsf{FV}(\varphi) \subseteq \mathsf{FV}(\psi)$ holds. So, if $\varphi \to \psi$ is an involvement, K-axiom is valid.

Proposition 1. *1.* $\Box\forall x.\varphi \to \forall x.\Box\varphi$ *is valid.*
2. If $\mathsf{FV}(\varphi) \subseteq \mathsf{FV}(\psi)$, *then* $\Box(\varphi \to \psi) \to (\Box\varphi \to \Box\psi)$ *is valid.*
3. If φ *is valid, then* $\Box\varphi$ *is valid.*
4. If $\mathsf{FV}(\varphi) = \mathsf{FV}(\psi)$ *and* $\varphi \leftrightarrow \psi$ *is valid, then* $\Box\varphi \leftrightarrow \Box\psi$ *is valid.*
5. $\Box\varphi \to \mathsf{O}_{(x,y,z)}\varphi$ *is valid.*
6. When $u \notin \{x, y, z\}$, $\mathsf{O}_{(x,y,z)}\forall u.\varphi \to \forall u.\mathsf{O}_{(x,y,z)}\varphi$ *is valid.*
7. If $\mathsf{FV}(\varphi) \subseteq \mathsf{FV}(\psi)$, $\mathsf{O}_{(x,y,z)}(\varphi \to \psi) \to (\mathsf{O}_{(x,y,z)}\varphi \to \mathsf{O}_{(x,y,z)}\psi)$ *is valid.*
8. If φ *is valid, then* $\mathsf{O}_{(x,y,x)}\varphi$ *is valid.*
9. If $\mathsf{FV}(\varphi) = \mathsf{FV}(\psi)$ *and* $\varphi \leftrightarrow \psi$ *is valid, then* $\mathsf{O}_{(x,y,z)}\varphi \leftrightarrow \mathsf{O}_{(x,y,z)}\psi$ *is valid.*

Proof. Item 4 is obtained from Items 2 and 3. Item 5 holds because $R_{\bigcirc}(\alpha(x), \alpha(y), \alpha(z)) \subseteq R_{\Box}$. Item 8 is proved by Items 3 and 5 and Item 9 is shown from Items 7 and 8. In what follows, we establish Items 6 and 7.

For Item 6, let $u \notin \{x, y, z\}$ and fix any model $M = (W, D, R_{\Box}, R_{\mathsf{O}}, V)$, any variable assignment α and any world $w \in W$ such that $\mathsf{O}_{(x,y,z)}\forall u.\varphi \to \forall u.\mathsf{O}_{(x,y,z)}\varphi$ is an α_w-formula. We assume $u \in \mathsf{FV}(\varphi)$ below (otherwise, $\forall u.$ is a "dummy" or vacuous quantifier and so it is easily seen to be valid). Suppose that $M, \alpha, w \models \mathsf{O}_{(x,y,z)}\forall u.\varphi$. To show $M, \alpha, w \models \forall u.\mathsf{O}_{(x,y,z)}\varphi$, let us fix any $d \in D(w)$. We show $M, \alpha(u|d), w \models \mathsf{O}_{(x,y,z)}\varphi$. Let us also fix any $v \in W$ such that $(w, v) \in R_{\mathsf{O}}(\alpha(u|d)(x), \alpha(u|d)(y), \alpha(u|d)(z))$ and $\alpha(u|d)[\mathsf{FV}(\varphi)] \subseteq D(v)$. We establish $M, \alpha(u|d), v \models \varphi$. By $u \notin \{x, y, z\}$,

$$R_{\mathsf{O}}(\alpha(u|d)(x), \alpha(u|d)(y), \alpha(u|d)(z)) = R_{\mathsf{O}}(\alpha(x), \alpha(y), \alpha(z)).$$

By $\mathsf{FV}(\forall u.\varphi) = \mathsf{FV}(\varphi) \setminus \{u\}$, it follows from $\alpha(u|d)[\mathsf{FV}(\varphi)] \subseteq D(v)$ that $\alpha[\mathsf{FV}(\forall u.\varphi)] \subseteq D(v)$. So, by our supposition of $M, \alpha, w \models \mathsf{O}_{(x,y,z)}\forall u.\varphi$, we have $M, \alpha, v \models \forall u.\varphi$. By $u \in \mathsf{FV}(\varphi)$ and $\alpha(u|d)[\mathsf{FV}(\varphi)] \subseteq D(v)$, we get $d \in D(v)$. It follows that $M, \alpha(u|d), v \models \varphi$, as required.

For Item 7, suppose that $\mathsf{FV}(\varphi) \subseteq \mathsf{FV}(\psi)$ and fix any model $M = (W, D, R_{\Box}, R_{\mathsf{O}}, V)$, any variable assignment α and any world $w \in W$ such that $\mathsf{O}_{(x,y,z)}(\varphi \to \psi) \to (\mathsf{O}_{(x,y,z)}\varphi \to \mathsf{O}_{(x,y,z)}\psi)$ is an α_w-formula. Suppose that $M, \alpha, w \models \mathsf{O}_{(x,y,z)}(\varphi \to \psi)$ and $M, \alpha, w \models \mathsf{O}_{(x,y,z)}\varphi$. To show that $M, \alpha, w \models \mathsf{O}_{(x,y,z)}\psi$, fix any world $v \in W$ such that $(w, v) \in R_{\mathsf{O}}(\alpha(x), \alpha(y), \alpha(z))$ and $\alpha[\mathsf{FV}(\psi)] \subseteq D(v)$. Our goal is to establish $M, \alpha, v \models \psi$. It follows from $\mathsf{FV}(\varphi) \subseteq \mathsf{FV}(\psi)$ that $\alpha[\mathsf{FV}(\varphi)] \subseteq \alpha[\mathsf{FV}(\psi)] \subseteq D(v)$ and $\alpha[\mathsf{FV}(\varphi \to \psi)] \subseteq D(v)$. By the supposition, we obtain $M, \alpha, v \models \varphi \to \psi$ and $M, \alpha, v \models \varphi$. Therefore, we conclude that $M, \alpha, v \models \psi$. □

Table 1. Hilbert System $\mathsf{H}(\mathbf{CTSDAL}^{=})$

`(A1)`	$\varphi \to (\psi \to \varphi)$
`(A2)`	$(\varphi \to (\psi \to \chi)) \to ((\varphi \to \psi) \to (\varphi \to \chi))$
`(A3)`	$(\neg\psi \to \neg\varphi) \to (\varphi \to \psi)$
`(A4)`	$\forall x.\varphi \to \varphi[y/x]$
`(MP)`	From φ and $\varphi \to \psi$, we may infer ψ.
`(R∀)`	From $\varphi \to \psi[z/x]$, we may infer $\varphi \to \forall x.\psi$, provided $z \notin \mathsf{FV}(\varphi \to \forall x.\psi)$.
`(Ref=)`	$x = x$
`(Repl=)`	$x = y \to (\varphi[x/z] \leftrightarrow \varphi[y/z])$
`(Kinv□)`	$\Box(\varphi \to \psi) \to (\Box\varphi \to \Box\psi)$ if $\mathsf{FV}(\varphi) \subseteq \mathsf{FV}(\psi)$
`(Nec□)`	From φ, we may infer $\Box\varphi$.
`(□ ≠)`	$x \neq y \to \Box(x \neq y)$
`(KinvO)`	$\mathsf{O}_{(x,y,z)}(\varphi \to \psi) \to (\mathsf{O}_{(x,y,z)}\varphi \to \mathsf{O}_{(x,y,z)}\psi)$ if $\mathsf{FV}(\varphi) \subseteq \mathsf{FV}(\psi)$
`(Mix)`	$\Box\varphi \to \mathsf{O}_{(x,y,z)}\varphi$

Table 1 provides all the axioms and inference rules of Hilbert system $\mathsf{H}(\mathbf{CTSDAL}^{=})$. Based on the axiomatization of the first-order logic with equality, we add all axioms and inference rules for common sense modal predicate logic by Seligman [9] and two additional axioms for $\mathsf{O}_{(x,y,z)}$. As regards the axiom (`Repl=`), note that we allow φ to be possibly non-atomic formulas. We define the notion of *theorem* as usual and use $\vdash \varphi$ to mean that φ is a theorem of $\mathsf{H}(\mathbf{CTSDAL}^{=})$.

Proposition 2. *The following hold in* $\mathsf{H}(\mathbf{CTSDAL}^{=})$*:*

(RE$\Box$) *If* $\vdash \varphi \leftrightarrow \psi$ *and* $\mathsf{FV}(\varphi) = \mathsf{FV}(\psi)$, *then* $\vdash \Box\varphi \leftrightarrow \Box\psi$.
(NecO) *If* $\vdash \varphi$, *then* $\vdash \mathsf{O}_{(x,y,z)}\varphi$.
(REO) *If* $\vdash \varphi \leftrightarrow \psi$ *and* $\mathsf{FV}(\varphi) = \mathsf{FV}(\psi)$, *then* $\vdash \mathsf{O}_{(x,y,z)}\varphi \leftrightarrow \mathsf{O}_{(x,y,z)}\psi$.
(CREO) *If* $\vdash \varphi \rightarrow (\psi_1 \leftrightarrow \psi_2)$ *and* $\mathsf{FV}(\varphi) \subseteq \mathsf{FV}(\psi_1) = \mathsf{FV}(\psi_2)$,
then $\vdash \mathsf{O}_{(x,y,z)}\varphi \rightarrow (\mathsf{O}_{(x,y,z)}\psi_1 \leftrightarrow \mathsf{O}_{(x,y,z)}\psi_2)$.
($\Box$ =) $\vdash x = y \rightarrow \Box(x = y)$.
(O =) $\vdash x = y \rightarrow \mathsf{O}_{(x_1,x_2,x_3)}(x = y)$.
(O $\neq$) $\vdash x \neq y \rightarrow \mathsf{O}_{(x_1,x_2,x_3)}(x \neq y)$.

Proof. (RE$\Box$) holds by (Nec$\Box$) and (Kinv$\Box$). (NecO) holds by (Nec$\Box$) and (Mix). We get both (REO) and (CREO) from (NecO) and (KinvO). (O $\neq$) is obtained from (Mix) and ($\Box$ $\neq$). Since (O =) is obtained from (Mix) and ($\Box$ =), we focus on ($\Box$ =) below. For ($\Box$ =), we proceed as follows. By (Ref=) and (Nec$\Box$), we get $\vdash \Box(x = x)$. By (Repl=), we have $\vdash x = y \rightarrow (\Box(x = x) \leftrightarrow \Box(x = y))$. By propositional reasoning, we get from $\vdash \Box(x = x)$ that $\vdash x = y \rightarrow \Box(x = y)$, as required. $\Box$

Remark 1. Note that the following analogue of the so-called Axiom (D) of standard deontic logic is not included in the proof system defined above:

$$\mathsf{O}_{(x,y,z)}\varphi \rightarrow \mathsf{P}_{(x,y,z)}\varphi.$$

Axiom (D) has the effect of precluding the possibility of deontic explosion (an absurd situation in which everything comes to be obligatory) by precluding the possibility of conflict of obligations in standard deontic logic (**SDL**, for short). As $\mathsf{H}(\mathbf{CTSDAL}^{=})$ will be dynamically extended into a dynamic logic of acts of commanding and promising in the next section, however, we need to keep open the possibility of conflicts of obligations being generated by acts of commanding and/or promising. This, of course, leaves the possibility of deontic explosion, but it will be shown to be minimized in the dynamic logic to be developed. In the next section, we will show how conflicts of obligations is generated and how the situation is represented without triggering deontic explosion in real life examples (see Example 1).

Apart from the exclusion of Axiom (D) and the inclusion of Axiom (Mix), which implies that if something is permitted then it is possible, $\mathsf{H}(\mathbf{CTSDAL}^{=})$ can be said to be similar to **SDL** in that it includes a weaker version (KinvO) of Axiom (K) for deontic operators and the necessitation rule (NecO) can be derived from (Nec$\Box$) and (Mix) as in Proposition 2.

Theorem 1. $\mathsf{H}(\mathbf{CTSDAL}^{=})$ *is sound for the semantics, i.e., if a formula is a theorem in it, then it is valid.*

Proof. It suffices to check that all the axioms are valid and all inference rules preserve validity. We only focus on the non-first order axioms and inference rules. By Proposition 1, it suffices to show the validity of ($\Box$ $\neq$). Fix any model $M = (W, D, R_{\Box}, R_{\mathsf{O}}, V)$, any variable assignment α and any world $w \in W$ such that $x \neq y \rightarrow \Box(x \neq y)$ is an α_w-formula. Suppose that $M, \alpha, w \models x \neq y$, i.e., $\alpha(x) \neq \alpha(y)$. Fix any world $v \in W$ such that $(w, v) \in R_{\Box}$ and $\alpha[\{x, y\}] \subseteq D(v)$. So, $x \neq y$ is an α_v-formula. By $\alpha(x) \neq \alpha(y)$, we conclude $M, \alpha, v \models x \neq y$, as desired. $\Box$

Remark 2. Semantic completeness of the fragment of H(**CTSDAL**$^{=}$) without deontic operators and equality was already established in [8] by constructing a canonical model. Moreover, Yamada [16] employed an argument in [8] to establish semantic completeness of the fragment of H(**CTSDAL**$^{=}$) without equality. But, it is still an open question if H(**CTSDAL**$^{=}$) is semantically complete, i.e., every valid formula φ is a theorem of H(**CTSDAL**$^{=}$).

3 Dynamification: Acts of Commanding and Promising

We extend our previous syntax $\mathcal{L}$ with dynamic operators of the form $[\mathsf{Com}_{(x,y)}\cdot]$ and $[\mathsf{Prom}_{(x,y)}\cdot]$ for acts of commanding and promising and denote this new syntax by $\mathcal{L}^{+}$. Then, the set $\mathsf{Form}_{\mathcal{L}^{+}}$ of all formulas in $\mathcal{L}^{+}$ is inductively defined as follows:

$$\mathsf{Form}_{\mathcal{L}^{+}} \ni \varphi ::= P(x_1, \ldots, x_n) \mid x = y \mid \neg\varphi \mid (\varphi \to \varphi) \mid \forall x.\varphi \mid \Box\varphi \mid \mathsf{O}_{(x,y,z)}\varphi \mid [\pi]\varphi,$$
$$\pi ::= \mathsf{Com}_{(x,y)}\varphi \mid \mathsf{Prom}_{(x,y)}\varphi.$$

So, we can construct formulas such as $[\mathsf{Com}_{(x,y)}\varphi]\psi$ and $[\mathsf{Prom}_{(x,y)}\varphi]\psi$. We also define the complexity $\mathsf{c}(\varphi)$ of a formula $\varphi \in \mathsf{Form}_{\mathcal{L}^{+}}$ as the total number of occurrences of logical connectives and dynamic operators in φ, e.g., $\mathsf{c}([\mathsf{Com}_{(x,y)}\varphi]\psi) = \mathsf{c}(\varphi) + \mathsf{c}(\psi) + 1$. For a formula $\varphi \in \mathsf{Form}_{\mathcal{L}^{+}}$, we also use $\mathsf{FV}(\varphi)$ to mean the set of all free variables in φ by adding the following clauses to Definition 1:

$$\mathsf{FV}([\mathsf{Com}_{(x,y)}\varphi]\psi) := \{x, y\} \cup \mathsf{FV}(\varphi) \cup \mathsf{FV}(\psi),$$
$$\mathsf{FV}([\mathsf{Prom}_{(x,y)}\varphi]\psi) := \{x, y\} \cup \mathsf{FV}(\varphi) \cup \mathsf{FV}(\psi).$$

We read $[\mathsf{Com}_{(x,y)}\varphi]\psi$ as "whenever an agent x commands an agent y to see to it that φ, ψ holds in the resulting situation". On the other hand, we read $[\mathsf{Prom}_{(x,y)}\varphi]\psi$ as "whenever an agent x promises an agent y that x will see to it that φ, ψ holds in the resulting situation". Semantically, the act $[\mathsf{Com}_{(x,y)}\varphi]$ of commanding restricts the range of the deontic accessibility relation for the triple (y, x, x) to φ-states. Similarly, the act $[\mathsf{Prom}_{(x,y)}\varphi]$ of promising restricts the range of the deontic accessibility relation for the triple (x, y, x) to φ-states. These ideas are implemented as in the following definition.

Definition 4. *Let $M = (W, D, R_{\Box}, R_{\mathsf{O}}, V)$ be a model and α be a variable assignment. We extend the* satisfaction relation *$M, \alpha, w \models \varphi$ of Definition 3 between M, α, w and α_w-formula φ by the following new clause:*

$$M, \alpha, w \models [\pi]\psi \text{ iff } M^{\pi}_{\alpha}, \alpha, w \models \psi,$$

where $M^{\pi}_{\alpha} = (W, D, R_{\Box}, R^{\pi}_{\mathsf{O}}, V)$ and $R^{\pi}_{\mathsf{O}} \subseteq W \times W$ is defined as follows.

- *Let $\pi := \mathsf{Com}_{(x,y)}\varphi$. If $(d_1, d_2, d_3) = (\alpha(y), \alpha(x), \alpha(x))$ then we define:*

$$(w, v) \in R^{\pi}_{\mathsf{O}}(d_1, d_2, d_3) \text{ iff } (w, v) \in R_{\mathsf{O}}(d_1, d_2, d_3) \text{ and } M, \alpha, v \models \varphi.$$

  Otherwise, $R^{\pi}_{\mathsf{O}}(d_1, d_2, d_3) := R_{\mathsf{O}}(d_1, d_2, d_3)$.

Table 2. Recursion Axioms for Act of Commanding

(CPr)	$[\mathsf{Com}_{(x,y)}\varphi]P(x_1,\dots,x_n)$	$\leftrightarrow$	$(\top_{\{x,y\}\cup\mathsf{FV}(\varphi)} \to P(x_1,\dots,x_n))$
(C=)	$[\mathsf{Com}_{(x,y)}\varphi]x_1 = x_2$	$\leftrightarrow$	$(\top_{\{x,y\}\cup\mathsf{FV}(\varphi)} \to x_1 = x_2)$
(C¬)	$[\mathsf{Com}_{(x,y)}\varphi]\neg\psi$	$\leftrightarrow$	$\neg[\mathsf{Com}_{(x,y)}\varphi]\psi$
(C →)	$[\mathsf{Com}_{(x,y)}\varphi](\psi_1 \to \psi_2)$	$\leftrightarrow$	$([\mathsf{Com}_{(x,y)}\varphi]\psi_1 \to [\mathsf{Com}_{(x,y)}\varphi]\psi_2)$
(C∀)	$[\mathsf{Com}_{(x,y)}\varphi]\forall z.\,\psi$	$\leftrightarrow$	$\forall u.\,([\mathsf{Com}_{(x,y)}\varphi]\psi[u/z])$ (u is fresh)
(C□)	$[\mathsf{Com}_{(x,y)}\varphi]\Box\psi$	$\leftrightarrow$	$\Box[\mathsf{Com}_{(x,y)}\varphi]\psi$
(CO)	$[\mathsf{Com}_{(x,y)}\varphi]\mathsf{O}_{(x_1,x_2,x_3)}\psi$	$\leftrightarrow$	$((x_1 = y) \wedge (x_2 = x) \wedge (x_3 = x) \to \mathsf{O}_{(x_1,x_2,x_3)}(\varphi \to [\mathsf{Com}_{(x,y)}\varphi]\psi))$ $\wedge(\neg((x_1 = y) \wedge (x_2 = x) \wedge (x_3 = x)) \to \mathsf{O}_{(x_1,x_2,x_3)}[\mathsf{Com}_{(x,y)}\varphi]\psi)$

Table 3. Recursion Axioms for Act of Promising

(PPr)	$[\mathsf{Prom}_{(x,y)}\varphi]P(x_1,\dots,x_n)$	$\leftrightarrow$	$(\top_{\{x,y\}\cup\mathsf{FV}(\varphi)} \to P(x_1,\dots,x_n))$
(P=)	$[\mathsf{Prom}_{(x,y)}\varphi]x_1 = x_2$	$\leftrightarrow$	$(\top_{\{x,y\}\cup\mathsf{FV}(\varphi)} \to x_1 = x_2)$
(P¬)	$[\mathsf{Prom}_{(x,y)}\varphi]\neg\psi$	$\leftrightarrow$	$\neg[\mathsf{Prom}_{(x,y)}\varphi]\psi$
(P →)	$[\mathsf{Prom}_{(x,y)}\varphi](\psi_1 \to \psi_2)$	$\leftrightarrow$	$([\mathsf{Prom}_{(x,y)}\varphi]\psi_1 \to [\mathsf{Prom}_{(x,y)}\varphi]\psi_2)$
(P∀)	$[\mathsf{Prom}_{(x,y)}\varphi]\forall z.\,\psi$	$\leftrightarrow$	$\forall u.\,([\mathsf{Prom}_{(x,y)}\varphi]\psi[u/z])$ (u is fresh)
(P□)	$[\mathsf{Prom}_{(x,y)}\varphi]\Box\psi$	$\leftrightarrow$	$\Box[\mathsf{Prom}_{(x,y)}\varphi]\psi$
(PO)	$[\mathsf{Prom}_{(x,y)}\varphi]\mathsf{O}_{(x_1,x_2,x_3)}\psi$	$\leftrightarrow$	$((x_1 = x) \wedge (x_2 = y) \wedge (x_3 = x) \to \mathsf{O}_{(x_1,x_2,x_3)}(\varphi \to [\mathsf{Prom}_{(x,y)}\varphi]\psi))$ $\wedge(\neg((x_1 = x) \wedge (x_2 = y) \wedge (x_3 = x)) \to \mathsf{O}_{(x_1,x_2,x_3)}[\mathsf{Prom}_{(x,y)}\varphi]\psi)$

– *Let* $\pi := \mathsf{Prom}_{(x,y)}\varphi$. *If* $(d_1, d_2, d_3) = (\alpha(x), \alpha(y), \alpha(x))$ *then we define:*

$$(w, v) \in R^{\pi}_{\mathsf{O}}(d_1, d_2, d_3) \text{ iff } (w, v) \in R_{\mathsf{O}}(d_1, d_2, d_3) \text{ and } M, \alpha, v \models \varphi.$$

Otherwise, $R^{\pi}_{\mathsf{O}}(d_1, d_2, d_3) := R_{\mathsf{O}}(d_1, d_2, d_3)$.

We define the notion of validity *as in Definition 3.*

Proposition 3. *1. All the formulas in Table 2 are valid.*
2. All the formulas in Table 3 are valid.

Proof. We only establish Item 1 because Item 2 is proved similarly. For Item 1, we show the validity of (CPr), (C∀) and (CO).

First, we deal with (CPr). Fix any model $M = (W, D, R_{\Box}, R_{\mathsf{O}}, V)$, any variable assignment α, and any world $w \in W$ such that every free variable x in (CPr) satisfies $\alpha(x) \in D(w)$. We proceed as follows:

$$\begin{aligned} M, \alpha, w \models [\mathsf{Com}_{(x,y)}\varphi]P(x_1, \dots, x_n) &\text{ iff } M_{\alpha}^{\mathsf{Com}_{(x,y)}\varphi}, \alpha, w \models P(x_1, \dots, x_n), \\ &\text{ iff } (\alpha(x_1), \dots, \alpha(x_n)) \in V(P, w), \\ &\text{ iff } M, \alpha, w \models P(x_1, \dots, x_n), \\ &\text{ iff } M, \alpha, w \models \top_{\{x,y\}\cup\mathsf{FV}(\varphi)} \to P(x_1, \dots, x_n), \end{aligned}$$

where we note that $\alpha[\{x, y\} \cup \mathsf{FV}(\varphi)] \subseteq D(w)$ hence $M, \alpha, w \models \top_{\{x,y\}\cup\mathsf{FV}(\varphi)}$.

Second, we deal with (C∀). Fix any model $M = (W, D, R_{\Box}, R_{\mathsf{O}}, V)$, any variable assignment α, and any world $w \in W$ such that every free variable x in (C∀) satisfies

$\alpha(x) \in D(w)$. We proceed as follows:

$$\begin{aligned}
M, \alpha, w \models [\mathsf{Com}_{(x,y)}\varphi]\forall z.\psi \text{ iff } & M_\alpha^{\mathsf{Com}_{(x,y)}\varphi}, \alpha, w \models \forall z.\psi \\
\text{iff } & M_\alpha^{\mathsf{Com}_{(x,y)}\varphi}, \alpha(z|d), w \models \psi, \text{ for every } d \in D(w), \\
\text{iff } & M_\alpha^{\mathsf{Com}_{(x,y)}\varphi}, \alpha(u|d), w \models \psi[u/z], \text{ for every } d \in D(w), \\
\text{iff } & M_{\alpha(u|d)}^{\mathsf{Com}_{(x,y)}\varphi}, \alpha(u|d), w \models \psi[u/z], \text{ for every } d \in D(w), \\
\text{iff } & M, \alpha, w \models \forall u.([\mathsf{Com}_{(x,y)}\varphi](\psi[u/z])),
\end{aligned}$$

where we note that $u \notin \{x, y\} \cup \mathsf{FV}(\psi)$ and so $M_{\alpha(u|d)}^{\mathsf{Com}_{(x,y)}\varphi} = M_\alpha^{\mathsf{Com}_{(x,y)}\varphi}$ by definition.

Finally, we deal with (CO). Fix any model $M = (W, D, R_\Box, R_\mathsf{O}, V)$, any variable assignment α, and any world $w \in W$ such that every free variables x in (CO) satisfies $\alpha(x) \in D(w)$. We proceed as follows:

$$\begin{aligned}
& M, \alpha, w \models [\mathsf{Com}_{(x,y)}\varphi]\mathsf{O}_{(x_1,x_2,x_3)}\psi \\
\text{iff } & M_\alpha^{\mathsf{Com}_{(x,y)}\varphi}, \alpha, w \models \mathsf{O}_{(x_1,x_2,x_3)}\psi, \\
\text{iff } & \text{For all } v, \text{ if } (w,v) \in R_\mathsf{O}^\pi(\alpha(x_1), \alpha(x_2), \alpha(x_3)) \text{ and } \alpha[\mathsf{FV}(\varphi)] \subseteq D(v), \text{ then } M_\alpha^{\mathsf{Com}_{(x,y)}\varphi}, \alpha, v \models \psi, \\
\text{iff } & \text{For all } v, \text{ if } (w,v) \in R_\mathsf{O}^\pi(\alpha(x_1), \alpha(x_2), \alpha(x_3)) \text{ and } \alpha[\mathsf{FV}(\varphi)] \subseteq D(v), \text{ then } M, \alpha, v \models [\mathsf{Com}_{(x,y)}\varphi]\psi.
\end{aligned}$$

where $\pi := \mathsf{Com}_{(x,y)}\varphi$. By the definition of R_O^π when $\pi := \mathsf{Com}_{(x,y)}\varphi$, the last line is equivalent to the conjunction of the following two statements:

- If $(\alpha(x_1), \alpha(x_2), \alpha(x_3)) = (\alpha(y), \alpha(x), \alpha(x))$, then, for all worlds $v \in W$,

  if $(w, v) \in R_\mathsf{O}(\alpha(x_1), \alpha(x_2), \alpha(x_3))$ and $M, \alpha, v \models \varphi$ and $\alpha[\mathsf{FV}(\varphi)] \subseteq D(v)$, then $M, \alpha, v \models [\mathsf{Com}_{(x,y)}\varphi]\psi$.

- Otherwise, for all worlds $v \in W$,

  if $(w, v) \in R_\mathsf{O}(\alpha(x_1), \alpha(x_2), \alpha(x_3))$ and $\alpha[\mathsf{FV}(\varphi)] \subseteq D(v)$, then $M, \alpha, v \models [\mathsf{Com}_{(x,y)}\varphi]\psi$.

The conjunction of these two statements is equivalent to the conjunction of the following:

- If $M, \alpha, w \models (x_1 = y) \wedge (x_2 = x) \wedge (x_3 = x)$, then

$$M, \alpha, w \models \mathsf{O}_{(x_1,x_2,x_3)}(\varphi \to [\mathsf{Com}_{(x,y)}\varphi]\psi).$$

- If $M, \alpha, w \models \neg((x_1 = y) \wedge (x_2 = x) \wedge (x_3 = x))$, then

$$M, \alpha, w \models \mathsf{O}_{(x_1,x_2,x_3)}[\mathsf{Com}_{(x,y)}\varphi]\psi.$$

This conjunction is easily seen to be equivalent to the right hand side of the axiom (CO) of Table 2. □

Proposition 4. *The following inference rules preserve validity:*

(NecC) *From* ψ, *we may infer* $[\mathsf{Com}_{(x,y)}\varphi]\psi$;
(NecP) *From* ψ, *we may infer* $[\mathsf{Prom}_{(x,y)}\varphi]\psi$.

Proof. As for (NecP), we proceed as follows. Assume that ψ is valid. To show that $[\mathsf{Prom}_{(x,y)}\varphi]\psi$ is valid, let us fix any model $M = (W, D, R_\Box, R_\mathsf{O}, V)$, any variable assignment α and any world $w \in W$ such that $[\mathsf{Prom}_{(x,y)}\varphi]\psi$ is an α_w-formula. Our goal is to show that $M_\alpha^{\mathsf{Prom}_{(x,y)}\varphi}, \alpha, w \models \psi$. Since ψ is valid and ψ is also an α_w-formula, we obtain $M_\alpha^{\mathsf{Prom}_{(x,y)}\varphi}, \alpha, w \models \psi$, as desired. We can also similarly establish that (NecC) preserves validity. □

Remark 3. It can be seen from the validity of (C →) and (P →) from Proposition 3 and (NecC) and (NecP) from Proposition 4 that both $[\mathsf{Com}_{(x,y)}\varphi]$ and $[\mathsf{Prom}_{(x,y)}\varphi]$ can be regarded as *normal* modal operators.

In what follows, we use the term *CGO* and *PGO Principles* to denote formulas of the form $\forall x. \forall y. [\mathsf{Com}_{(x,y)}\varphi]\mathsf{O}_{(y,x,x)}\varphi$ and $\forall x. \forall y. [\mathsf{Prom}_{(x,y)}\varphi]\mathsf{O}_{(x,y,x)}\varphi$, respectively, where "CGO" and "PGO" abbreviate "act of commanding generates obligation" and "act of promising generates obligation", respectively. We remark that the notion of CGO and PGO Principles are similar to the notion of successful formulas in the public announcement logic (cf. [11, Definition 4.31] and see also [6]).

Definition 5. *We say that* CGO (or PGO) Principle holds for $\Gamma \subseteq \mathsf{Form}_{\mathcal{L}^+}$ *if, for every* $\varphi \in \Gamma$, $\forall x. \forall y. [\mathsf{Com}_{(x,y)}\varphi]\mathsf{O}_{(y,x,x)}\varphi$ *(or* $\forall x. \forall y. [\mathsf{Prom}_{(x,y)}\varphi]\mathsf{O}_{(x,y,x)}\varphi$*, respectively) is valid.*

We are going to specify at least two classes of formulas from $\mathsf{Form}_{\mathcal{L}}$ for which CGO and PGO Principles hold.

Lemma 1. *1. If* $\varphi \in \mathsf{Form}_{\mathcal{L}}$ *is a first-order formula (i.e., it does not contain* $\Box$ *and* $\mathsf{O}_{(x_1,x_2,x_3)}$*), then* $\varphi \leftrightarrow [\mathsf{Com}_{(x,y)}\psi]\varphi$ *and* $\varphi \leftrightarrow [\mathsf{Prom}_{(x,y)}\psi]\varphi$ *are valid.*
2. If $\varphi \in \mathsf{Form}_{\mathcal{L}}$ *does not contain any deontic operator and* $\{x, y\} \cup \mathsf{FV}(\psi) \subseteq \mathsf{FV}(\varphi)$*, then* $\varphi \leftrightarrow [\mathsf{Com}_{(x,y)}\psi]\varphi$ *and* $\varphi \leftrightarrow [\mathsf{Prom}_{(x,y)}\psi]\varphi$ *are valid.*

Proof. We only show Item 2 alone because an argument for Item 1 can be obtained by omitting the assumption $\{x, y\} \cup \mathsf{FV}(\psi) \subseteq \mathsf{FV}(\varphi)$. For Item 2, we only prove the validity of $\varphi \leftrightarrow [\mathsf{Com}_{(x,y)}\psi]\varphi$. We argue by induction on $\mathsf{c}(\varphi)$ such that $\varphi \in \mathsf{Form}_{\mathcal{L}}$ does not contain any deontic operator and $\{x, y\} \cup \mathsf{FV}(\psi) \subseteq \mathsf{FV}(\varphi)$. We only present the cases where φ is of the form $P(x_1, \ldots, x_n)$, $\forall z. \chi$ or $\Box\chi$.

Let φ be $P(x_1, \ldots, x_n)$ such that $\{x, y\} \cup \mathsf{FV}(\psi) \subseteq \{x_1, \ldots, x_n\}$. We prove that $P(x_1, \ldots, x_n) \leftrightarrow [\mathsf{Com}_{(x,y)}\psi]P(x_1, \ldots, x_n)$ is valid. By Proposition 3, the axiom (CPr) $(\top_{\{x,y\}\cup\mathsf{FV}(\psi)} \to P(x_1, \ldots, x_n)) \leftrightarrow [\mathsf{Com}_{(x,y)}\psi]P(x_1, \ldots, x_n)$ is valid. Since $\top_{\{x,y\}\cup\mathsf{FV}(\psi)}$ is valid, our goal follows immediately.

Let φ be $\forall z. \chi$ such that $\forall z. \chi$ does not contain any deontic operator and $\{x, y\} \cup \mathsf{FV}(\psi) \subseteq \mathsf{FV}(\forall z. \chi) = \mathsf{FV}(\chi) \setminus \{z\}$. We prove that $\forall z. \chi \leftrightarrow [\mathsf{Com}_{(x,y)}\psi]\forall z. \chi$ is valid. By Proposition 3, the axiom (C∀) $\forall u. ([\mathsf{Com}_{(x,y)}\psi](\chi[u/z])) \leftrightarrow [\mathsf{Com}_{(x,y)}\psi]\forall z. \chi$ is valid, where u is a fresh variable. Since u is fresh, we have $\{x, y\} \cup \mathsf{FV}(\psi) \subseteq \mathsf{FV}(\chi[u/z])$. By induction hypothesis, we have the validity of $\chi[u/z] \leftrightarrow [\mathsf{Com}_{(x,y)}\psi](\chi[u/z])$. By the first-order reasoning, we obtain the validity of $\forall z. \chi \leftrightarrow \forall u. [\mathsf{Com}_{(x,y)}\psi](\chi[u/z])$. By the validity of (C∀) above, we conclude the desired validity of the equivalence $\forall z. \chi \leftrightarrow [\mathsf{Com}_{(x,y)}\psi]\forall z. \chi$.

Let φ be $\Box\chi$ such that $\Box\chi$ does not contain any deontic operator and $\{x, y\} \cup \mathsf{FV}(\psi) \subseteq \mathsf{FV}(\Box\chi) = \mathsf{FV}(\chi)$. We prove that $\Box\chi \leftrightarrow [\mathsf{Com}_{(x,y)}\psi]\Box\chi$ is valid. By induction hypothesis, $\chi \leftrightarrow [\mathsf{Com}_{(x,y)}\psi]\chi$ is valid. By our assumption of $\{x, y\} \cup \mathsf{FV}(\psi) \subseteq \mathsf{FV}(\chi)$, we get $\mathsf{FV}(\chi) = \mathsf{FV}([\mathsf{Com}_{(x,y)}\psi]\chi)$. It follows from Proposition 2 that $\Box\chi \leftrightarrow \Box[\mathsf{Com}_{(x,y)}\psi]\chi$ is valid. By Proposition 3, the validity of (C$\Box$) implies our desired goal, i.e., the validity of $\Box\chi \leftrightarrow [\mathsf{Com}_{(x,y)}\psi]\Box\chi$. □

Theorem 2. *Let $\varphi \in \mathsf{Form}_{\mathcal{L}}$. If* (i) *$\varphi$ is a first-order formula or* (ii) *φ does not contain any deontic operator and φ satisfies $\{ x, y \} \subseteq \mathsf{FV}(\varphi)$, then $\forall x. \forall y. [\mathsf{Com}_{(x,y)}\varphi]\mathsf{O}_{(y,x,x)}\varphi$ and $\forall x. \forall y. [\mathsf{Prom}_{(x,y)}\varphi]\mathsf{O}_{(x,y,x)}\varphi$ are valid.*

Proof. We only establish that the assumption (ii) implies our desired goal, because almost the same argument can be applied for (i). Assume that $\varphi \in \mathsf{Form}_{\mathcal{L}}$ does not contain any deontic operator and suppose that $\{ x, y \} \subseteq \mathsf{FV}(\varphi)$. We only prove the validity of $\forall x. \forall y. [\mathsf{Com}_{(x,y)}\varphi]\mathsf{O}_{(y,x,x)}\varphi$. Fix any model M, any variable assignment α and any world w in M. We show that $M, \alpha, w \models \forall x. \forall y. [\mathsf{Com}_{(x,y)}\varphi]\mathsf{O}_{(y,x,x)}\varphi$. Fix any elements $d, e \in D(w)$. We establish $M, (\alpha(x|d))(y|e), w \models [\mathsf{Com}_{(x,y)}\varphi]\mathsf{O}_{(y,x,x)}\varphi$. By the validity of (CO), it suffices to show $M, (\alpha(x|d))(y|e), w \models \mathsf{O}_{(y,x,x)}(\varphi \rightarrow [\mathsf{Com}_{(x,y)}\varphi]\varphi)$. By Lemma 1, $\varphi \rightarrow [\mathsf{Com}_{(x,y)}\varphi]\varphi$ is valid. It follows from the necessitation law for $\mathsf{O}_{(y,x,x)}$ that $\mathsf{O}_{(y,x,x)}(\varphi \rightarrow [\mathsf{Com}_{(x,y)}\varphi]\varphi)$ is valid. This implies our goal immediately. □

Theorem 2 tells us that CGO and PGO principles holds when $\varphi \in \mathsf{Form}_{\mathcal{L}}$ is a first-order formula or when $\varphi \in \mathsf{Form}_{\mathcal{L}}$ does not contain any deontic operator and $\{ x, y \} \subseteq \mathsf{FV}(\varphi)$. Thus, as an application of Theorem 2, the following is an valid instance of PGO and CGO, respectively:

$$\forall x. \forall y. [\mathsf{Prom}_{(x,y)}\mathsf{Safe}(y)]\mathsf{O}_{(x,y,x)}\mathsf{Safe}(y),$$
$$\forall x. \forall y. [\mathsf{Com}_{(x,y)}\mathsf{Safe}(y)]\mathsf{O}_{(y,x,x)}\mathsf{Safe}(y),$$

since Safe(y) is a first-order formula.

CGO and PGO Principles enable us to describe conflicts of obligations without triggering deontic explosion. Consider the following scenario.

Example 1. (cf. [13, p.96]) Suppose you receive a letter from your political guru, in which she commands you to join an important political demonstration to be held in Tokyo next month. Unfortunately, it is to be held on the very same day on which an international one-day conference on logic is to be held in São Paulo, and you have already promised your former student who organizes that conference that you will give an invited lecture there. The conference is planned to be held in person and neither pre-recorded video lectures nor on-line lectures are accepted. It is possible for you to join the demonstration in Tokyo, but if you choose to do so, you will fail to keep your promise. It is also possible for you to give a lecture at the conference in São Paulo, but if you choose to do so, you will fail to obey your guru's command. No available means of transportation are fast enough to enable you to join both events on the same day even though the time in São Paulo is 12 h behind the time in Tokyo. You have to decide which alternative to choose.

In this scenario, your guru's command is in conflict with your earlier promise. Let a possible world w of a model N represent the situation before your promise, and take an assignment α such that $\alpha(x)$ is your guru, $\alpha(y)$ is the student and $\alpha(z)$ is you. In addition, let $F(z)$ and $G(z)$ be understood as meaning that z attends the conference in São Paulo and that z join the demonstration in Tokyo. We further assume that all of x, y and z are pairwise distinct in the world w of N. Then in the situation before your promise, the following holds:

$$[\mathsf{Prom}_{(z,y)}F(z)][\mathsf{Com}_{(x,z)}G(z)]((\mathsf{O}_{(z,y,z)}F(z) \wedge \mathsf{O}_{(z,x,x)}G(z))).$$

In what follows, let us establish that this holds in the world w of N. First of all, we have the following, where the 1st item is a corollary of Theorem 2.

Proposition 5. *The following are valid:*

1. $[\mathsf{Prom}_{(z,y)}F(z)][\mathsf{Com}_{(x,z)}G(z)]\mathsf{O}_{(z,x,x)}G(z)$,
2. $\neg((y = x) \wedge (z = x)) \to [\mathsf{Prom}_{(z,y)}F(z)][\mathsf{Com}_{(x,z)}G(z)]\mathsf{O}_{(z,y,z)}F(z)$.

Proof. For Item 1, we proceed as follows. By Theorem 2, $[\mathsf{Com}_{(x,z)}G(z)]\mathsf{O}_{(z,x,x)}G(z)$ is valid. By Proposition 4, $[\mathsf{Prom}_{(z,y)}F(z)][\mathsf{Com}_{(x,z)}G(z)]\mathsf{O}_{(z,x,x)}G(z)$ is also valid, as desired.

For Item 2, let us fix any model $M = (W, D, R_{\Box}, R_{\mathsf{O}}, V)$, any variable assignment α and any world $w \in W$ such that the formula in the item is an α_w-formula. Assume that $M, \alpha, w \models \neg((y = x) \wedge (z = x))$. This means: it is not the case that both $\alpha(y) = \alpha(x)$ and $\alpha(z) = \alpha(x)$ hold. We show $M, \alpha, w \models [\mathsf{Prom}_{(z,y)}F(z)][\mathsf{Com}_{(x,z)}G(z)]\mathsf{O}_{(z,y,z)}F(z)$. By putting $\pi_F := \mathsf{Prom}_{(z,y)}F(z)$ and $\pi_G := \mathsf{Com}_{(x,z)}G(z)$, it suffices to prove that $(M^{\pi_F}_{\alpha})^{\pi_G}_{\alpha}, \alpha, w \models \mathsf{O}_{(z,y,z)}F(z)$. Let us denote by $R^{\pi_F;\pi_G}_{\mathsf{O}}$ the deontic accessibility relations in the resulting model $(M^{\pi_F}_{\alpha})^{\pi_G}_{\alpha}$. Let us fix any world v such that (w, v) belongs to the deontic accessibility relation for $R^{\pi_F;\pi_G}_{\mathsf{O}}(\alpha(z), \alpha(y), \alpha(z))$. By assumption on x, y and z, we obtain: $(w, v) \in R_{\mathsf{O}}(\alpha(z), \alpha(y), \alpha(z))$ and $\alpha(z) \in V(F, v)$. Our goal is to show that $(M^{\pi_F}_{\alpha})^{\pi_G}_{\alpha}, \alpha, v \models F(z)$, i.e., $\alpha(z) \in V(F, v)$ since the acts of commanding and promising do not change a given valuation. But, this is already obtained. □

By Proposition 5 and our assumption on x, y, z at w of N, we conclude that our formalization above holds at w of N, because both $[\mathsf{Prom}_{(z,y)}F(z)]$ and $[\mathsf{Com}_{(x,z)}G(z)]$ are normal modal operators (as in Remark 3) and so the bundled operator "$[\mathsf{Prom}_{(z,y)}F(z)][\mathsf{Com}_{(x,z)}G(z)]$" commutes with the conjunction.

Since if you join the demonstration in Tokyo, you cannot attend the conference in São Paulo (this means that $\Box(\neg(F(z) \wedge G(z)))$ holds at w of N), we may also be said to have the following:

$$[\mathsf{Prom}_{(z,y)}F(z)][\mathsf{Com}_{(x,z)}G(z)]((\mathsf{O}_{(z,y,z)}F(z) \wedge \mathsf{O}_{(z,x,x)}\neg F(z))),$$

which is easily shown to hold at w of N.

Thus, **DCTSDAL$^{=}$** can show how an act of promising and an act of commanding comes to be in conflict in concrete cases. Note, however, that this conflict of obligations does not lead to deontic explosion as $\mathsf{O}_{(z,y,z)}$ and $\mathsf{O}_{(z,x,x)}$ are distinct operators and $\neg((y = z) \wedge (z = x))$ holds. And similarly, the following situations do not do so either if $\neg(x = y)$ holds.

1. $N, \alpha, w \models [\mathsf{Com}_{(x,z)}G(z)][\mathsf{Com}_{(y,z)}\neg G(z)](\mathsf{O}_{(z,x,x)}G(z) \wedge \mathsf{O}_{(z,y,y)}\neg G(z))$.
2. $N, \alpha, w \models [\mathsf{Prom}_{(z,y)}F(z)][\mathsf{Prom}_{(z,x)}\neg F(z)](\mathsf{O}_{(z,y,z)}F(z) \wedge \mathsf{O}_{(z,x,z)}\neg F(z))$.

By contrast, the following cases lead to deontic explosion.

3. $N, \alpha, w \models [\mathsf{Com}_{(x,z)}G(z)][\mathsf{Com}_{(x,z)}\neg G(z)](\mathsf{O}_{(z,x,x)}G(z) \wedge \mathsf{O}_{(z,x,x)}\neg G(z))$.
4. $N, \alpha, w \models [\mathsf{Prom}_{(z,y)}F(z)][\mathsf{Prom}_{(z,y)}\neg F(z)](\mathsf{O}_{(z,y,z)}F(z) \wedge \mathsf{O}_{(z,y,z)}\neg F(z))$.

Note that the deontic accessibility relation for the triple $(\alpha(z), \alpha(x), \alpha(x))$, say, of the updated model $(N_\alpha^{\mathsf{Com}_{(x,z)}G(z)})_\alpha^{\mathsf{Com}_{(x,z)}\neg G(z)}$ is an empty set, since the first update by $[\mathsf{Com}_{(x,z)}G(z)]$ cuts all the links to the worlds where $G(z)$ does not hold and the second update by $[\mathsf{Com}_{(x,z)}\neg G(z)]$ cuts all the links to the worlds where $\neg G(z)$ does not hold. Thus, for any ψ such that ψ is an α_w-formula, we vacuously have:

$$N, \alpha, w \models [\mathsf{Com}_{(x,z)}G(z)][\mathsf{Com}_{(x,z)}\neg G(z)]\mathsf{O}_{(z,x,x)}\psi.$$

Issuing such a pair of commands (or issuing a command of the form $[\mathsf{Com}_{(x,y)}(G(z) \wedge \neg G(z))]$) is irrational and should be avoided. A similar remark applies to acts of promising as well.

Remark 4. As is mentioned in Remark 1, the possibility of deontic explosion is minimized in H(**DCTSDAL**$^=$). Deontic explosion occurs only when one and the same agent gives command(s) (or promise(s)) with contradictory content(s) to one and the same addressee.

4 Relative Completeness

Definition 6. Hilbert system H(**DCTSDAL**$^=$) *is the extension of* H(**CTSDAL**$^=$) *by all the axioms of Tables 2 and 3 and the inference rules* (NecC) *and* (NecP) *as given in Proposition 4. We define the notion of theorem as usual and use* $\vdash^+ \varphi$ *to mean that* φ *is a theorem of* H(**DCTSDAL**$^=$).

Theorem 3. H(**DCTSDAL**$^=$) *is sound, i.e., for every formula* $\varphi \in \mathsf{Form}_{\mathcal{L}^+}$, *if* φ *is a theorem of* H(**DCTSDAL**$^=$) *then* φ *is valid.*

Proof. In addition to our argument for Theorem 1, Proposition 3 and Proposition 4 tell us that all the remaining axioms are valid and all the remaining rules preserve validity. □

Definition 7. *A translation* $t : \mathsf{Form}_{\mathcal{L}^+} \to \mathsf{Form}_{\mathcal{L}}$ *is defined inductively as in Table 4.*

Definition 8. *Given a formula* φ, *we define the weight* $\mathtt{w}(\varphi)$ *of the formula* φ *as the number of all occurrences of dynamic operators in* φ *inductively as follows:*

$$\begin{aligned}
\mathtt{w}(P(x_1, \ldots, x_n)) &:= 0,\\
\mathtt{w}(z_1 = z_n) &:= 0,\\
\mathtt{w}(\neg\psi) &:= \mathtt{w}(\psi),\\
\mathtt{w}(\psi_1 \to \psi_2) &:= \mathtt{w}(\psi_1) + \mathtt{w}(\psi_2),\\
\mathtt{w}(\forall z.\, \psi) &:= \mathtt{w}(\psi),\\
\mathtt{w}(\Box\psi) &:= \mathtt{w}(\psi),\\
\mathtt{w}(\mathsf{O}_{(x,y,z)}\psi) &:= \mathtt{w}(\psi),\\
\mathtt{w}([\mathsf{Com}_{(x,y)}\varphi]\psi) &:= \mathtt{w}(\varphi) + \mathtt{w}(\psi) + 1,\\
\mathtt{w}([\mathsf{Prom}_{(x,y)}\varphi]\psi) &:= \mathtt{w}(\varphi) + \mathtt{w}(\psi) + 1.
\end{aligned}$$

Table 4. Translation from $\mathsf{Form}_{\mathcal{L}^+}$ to $\mathsf{Form}_{\mathcal{L}}$

$$
\begin{array}{ll}
t(P(x_1,\ldots,x_n)) & := P(x_1,\ldots,x_n),\\
t(z_1 = z_2) & := z_1 = z_2,\\
t(\neg\psi) & := \neg t(\psi),\\
t(\psi_1 \to \psi_2) & := t(\psi_1) \to t(\psi_2),\\
t(\forall z.\,\psi) & := \forall z.\, t(\psi),\\
t(\Box\psi) & := \Box t(\psi),\\
t(\mathsf{O}_{(x_1,x_2,x_3)}\psi) & := \mathsf{O}_{(x_1,x_2,x_3)} t(\psi),\\
t([\mathsf{Com}_{(x,y)}\varphi]P(x_1,\ldots,x_n)) & := \top_{\{x,y\}\cup\mathsf{FV}(\varphi)} \to P(x_1,\ldots,x_n),\\
t([\mathsf{Com}_{(x,y)}\varphi]z_1 = z_2) & := \top_{\{x,y\}\cup\mathsf{FV}(\varphi)} \to z_1 = z_2,\\
t([\mathsf{Com}_{(x,y)}\varphi]\neg\psi) & := \neg t([\mathsf{Com}_{(x,y)}\varphi]\psi),\\
t([\mathsf{Com}_{(x,y)}\varphi](\psi_1 \to \psi_2)) & := t([\mathsf{Com}_{(x,y)}\varphi]\psi_1) \to t([\mathsf{Com}_{(x,y)}\varphi]\psi_2),\\
t([\mathsf{Com}_{(x,y)}\varphi]\forall z.\,\psi) & := \forall u.\, t([\mathsf{Com}_{(x,y)}\varphi](\psi[u/z])) \text{ where } u \text{ is fresh},\\
t([\mathsf{Com}_{(x,y)}\varphi]\Box\psi) & := \Box t([\mathsf{Com}_{(x,y)}\varphi]\psi),\\
t([\mathsf{Com}_{(x,y)}\varphi]\mathsf{O}_{(x_1,x_2,x_3)}\psi) & := ((x_1 = y) \wedge (x_2 = x) \wedge (x_3 = x) \to \mathsf{O}_{(x_1,x_2,x_3)}(t(\varphi) \to t([\mathsf{Com}_{(x,y)}\varphi]\psi)))\\
& \quad \wedge(\neg((x_1 = y) \wedge (x_2 = x) \wedge (x_3 = x)) \to \mathsf{O}_{(x_1,x_2,x_3)} t([\mathsf{Com}_{(x,y)}\varphi]\psi)),\\
t([\mathsf{Com}_{(x,y)}\varphi][\mathsf{Com}_{(z,u)}\chi]\psi) & := t([\mathsf{Com}_{(x,y)}\varphi]t([\mathsf{Com}_{(z,u)}\chi]\psi)),\\
t([\mathsf{Com}_{(x,y)}\varphi][\mathsf{Prom}_{(z,u)}\chi]\psi) & := t([\mathsf{Com}_{(x,y)}\varphi]t([\mathsf{Prom}_{(z,u)}\chi]\psi)),\\
t([\mathsf{Prom}_{(x,y)}\varphi]P(x_1,\ldots,x_n)) & := \top_{\{x,y\}\cup\mathsf{FV}(\varphi)} \to P(x_1,\ldots,x_n),\\
t([\mathsf{Prom}_{(x,y)}\varphi]z_1 = z_2) & := \top_{\{x,y\}\cup\mathsf{FV}(\varphi)} \to z_1 = z_2,\\
t([\mathsf{Prom}_{(x,y)}\varphi]\neg\psi) & := \neg t([\mathsf{Prom}_{(x,y)}\varphi]\psi),\\
t([\mathsf{Prom}_{(x,y)}\varphi](\psi_1 \to \psi_2)) & := t([\mathsf{Prom}_{(x,y)}\varphi]\psi_1) \to t([\mathsf{Prom}_{(x,y)}\varphi]\psi_2),\\
t([\mathsf{Prom}_{(x,y)}\varphi]\forall z.\,\psi) & := \forall u.\, t([\mathsf{Prom}_{(x,y)}\varphi](\psi[u/z])) \text{ where } u \text{ is fresh},\\
t([\mathsf{Prom}_{(x,y)}\varphi]\Box\psi) & := \Box t([\mathsf{Prom}_{(x,y)}\varphi]\psi),\\
t([\mathsf{Prom}_{(x,y)}\varphi]\mathsf{O}_{(x_1,x_2,x_3)}\psi) & := ((x_1 = x) \wedge (x_2 = y) \wedge (x_3 = x) \to \mathsf{O}_{(x_1,x_2,x_3)}(t(\varphi) \to t([\mathsf{Prom}_{(x,y)}\varphi]\psi)))\\
& \quad \wedge(\neg((x_1 = x) \wedge (x_2 = y) \wedge (x_3 = x)) \to \mathsf{O}_{(x_1,x_2,x_3)} t([\mathsf{Prom}_{(x,y)}\varphi]\psi)),\\
t([\mathsf{Prom}_{(x,y)}\varphi][\mathsf{Com}_{(z,u)}\chi]\psi) & := t([\mathsf{Prom}_{(x,y)}\varphi]t([\mathsf{Com}_{(z,u)}\chi]\psi)),\\
t([\mathsf{Prom}_{(x,y)}\varphi][\mathsf{Prom}_{(z,u)}\chi]\psi) & := t([\mathsf{Prom}_{(x,y)}\varphi]t([\mathsf{Prom}_{(z,u)}\chi]\psi)).
\end{array}
$$

Lemma 2. *Let $\eta \in \mathsf{Form}_{\mathcal{L}}$, i.e., η does not contain any dynamic operators. Then, $t(\eta) \in \mathsf{Form}_{\mathcal{L}}$, $\mathsf{FV}(\eta) = \mathsf{FV}(t(\eta))$ and the equivalence $\eta \leftrightarrow t(\eta)$ is a theorem of* H(**DCTSDAL**$^=$).

Proof. By induction on $\mathsf{c}(\eta)$, we can show that $t(\eta)$ is an alphabetic variant of η, i.e., $t(\eta)$ is obtained by renaming some of the bound variables in η. Then, all statements are immediate. □

The following is crucial for applying replacement of equivalent formulas in H(**DCTSDAL**$^=$).

Lemma 3. *For any formula $\eta \in \mathsf{Form}_{\mathcal{L}^+}$, $t(\eta) \in \mathsf{Form}_{\mathcal{L}}$ and $\mathsf{FV}(\eta) = \mathsf{FV}(t(\eta))$.*

Proof. Because we add $\top_{\{x,y\}\cup\mathsf{FV}(\varphi)}$ for our recursion axioms for atomic formulas in Tables 2 and 3, we can obtain, for example,

$$
\begin{aligned}
\mathsf{FV}([\mathsf{Com}_{(x,y)}\varphi]P(x_1,\ldots,x_n)) &= \{x,y\} \cup \mathsf{FV}(\varphi) \cup \{x_1,\ldots,x_n\}\\
&= \mathsf{FV}(\top_{\{x,y\}\cup\mathsf{FV}(\varphi)} \to P(x_1,\ldots,x_n)).
\end{aligned}
$$

To be more precise, we establish the statements (i) $t(\eta) \in \mathsf{Form}_{\mathcal{L}}$ and (ii) $\mathsf{FV}(\eta) = \mathsf{FV}(t(\eta))$ simultaneously by induction on $n = \mathtt{w}(\eta)$. When $n = 0$, both (i) and (ii) are obtained by Lemma 2.

Let $n > 0$ and suppose that both (i) and (ii) hold for every formula η such that $\mathtt{w}(\eta) \leqslant n$. We are going to prove that both (i) and (ii) hold for every formula η such that $\mathtt{w}(\eta) \leqslant n + 1$. We show both items by induction on the complexity $\mathtt{c}(\eta)$ (this is the second induction or subinduction). In what follows, we only deal with the cases where $\eta = [\mathsf{Com}_{(x,y)}\varphi]P(x_1, \ldots, x_n)$, $[\mathsf{Com}_{(x,y)}\varphi](\psi_1 \to \psi_2)$, or $[\mathsf{Com}_{(x,y)}\varphi][\mathsf{Prom}_{(z,u)}\chi]\psi$.

- Let η be of the form $[\mathsf{Com}_{(x,y)}\varphi]P(x_1, \ldots, x_n)$. (i) is immediate and (ii) is obvious from our initial observation in this proof.
- Let η be of the form $[\mathsf{Com}_{(x,y)}\varphi](\psi_1 \to \psi_2)$. As for (i), $t([\mathsf{Com}_{(x,y)}\varphi](\psi_1 \to \psi_2)) = t([\mathsf{Com}_{(x,y)}\varphi]\psi_1) \to t([\mathsf{Com}_{(x,y)}\varphi]\psi_2) \in \mathsf{Form}_{\mathcal{L}}$ by induction hypothesis of the second induction. For (ii), we proceed as follows:

$$\begin{aligned}\mathsf{FV}([\mathsf{Com}_{(x,y)}\varphi](\psi_1 \to \psi_2)) &= \mathsf{FV}([\mathsf{Com}_{(x,y)}\varphi]\psi_1) \cup \mathsf{FV}([\mathsf{Com}_{(x,y)}\varphi]\psi_2)\\ &= \mathsf{FV}(t([\mathsf{Com}_{(x,y)}\varphi]\psi_1)) \cup \mathsf{FV}(t([\mathsf{Com}_{(x,y)}\varphi]\psi_2))\\ &= \mathsf{FV}(t([\mathsf{Com}_{(x,y)}\varphi](\psi_1 \to \psi_2))),\end{aligned}$$

  where the second last equality is due to induction hypothesis of the second induction.
- Let η be of the form $[\mathsf{Com}_{(x,y)}\varphi][\mathsf{Prom}_{(z,u)}\chi]\psi$. First, we establish (i). By induction hypothesis of the second induction, we know that $t([\mathsf{Prom}_{(z,u)}\chi]\psi) \in \mathsf{Form}_{\mathcal{L}}$. Because $\mathtt{w}([\mathsf{Com}_{(x,y)}\varphi]t([\mathsf{Prom}_{(z,u)}\chi]\psi)) < \mathtt{w}([\mathsf{Com}_{(x,y)}\varphi][\mathsf{Prom}_{(z,u)}\chi]) \leqslant n + 1$, we obtain $\mathtt{w}([\mathsf{Com}_{(x,y)}\varphi]t([\mathsf{Prom}_{(z,u)}\chi]\psi)) \leqslant n$. Thus, it follows from induction hypothesis of the first induction (on weight n) that $t([\mathsf{Com}_{(x,y)}\varphi]t([\mathsf{Prom}_{(z,u)}\chi]\psi)) \in \mathsf{Form}_{\mathcal{L}}$. Second, we show (ii). By induction hypothesis of the second induction, we have $\mathsf{FV}([\mathsf{Prom}_{(z,u)}\chi]\psi) = \mathsf{FV}(t([\mathsf{Prom}_{(z,u)}\chi]\psi))$. We proceed as follows:

$$\begin{aligned}\mathsf{FV}([\mathsf{Com}_{(x,y)}\varphi][\mathsf{Prom}_{(z,u)}\chi]\psi) &= \{\, x, y \,\} \cup \mathsf{FV}(\varphi) \cup \mathsf{FV}([\mathsf{Prom}_{(z,u)}\chi]\psi)\\ &= \{\, x, y \,\} \cup \mathsf{FV}(\varphi) \cup \mathsf{FV}(t([\mathsf{Prom}_{(z,u)}\chi]\psi))\\ &= \mathsf{FV}([\mathsf{Com}_{(x,y)}\varphi]t([\mathsf{Prom}_{(z,u)}\chi]\psi)).\end{aligned}$$

  Since $\mathtt{w}([\mathsf{Com}_{(x,y)}\varphi]t([\mathsf{Prom}_{(z,u)}\chi]\psi)) < \mathtt{w}([\mathsf{Com}_{(x,y)}\varphi][\mathsf{Prom}_{(z,u)}\chi]) \leqslant n+1$, we have $\mathtt{w}([\mathsf{Com}_{(x,y)}\varphi]t([\mathsf{Prom}_{(z,u)}\chi]\psi)) \leqslant n$. By induction hypothesis of the first induction, we obtain

$$\begin{aligned}\mathsf{FV}([\mathsf{Com}_{(x,y)}\varphi]t([\mathsf{Prom}_{(z,u)}\chi]\psi)) &= \mathsf{FV}(t([\mathsf{Com}_{(x,y)}\varphi]t([\mathsf{Prom}_{(z,u)}\chi]\psi)))\\ &= \mathsf{FV}(t([\mathsf{Com}_{(x,y)}\varphi][\mathsf{Prom}_{(z,u)}\chi]\psi)),\end{aligned}$$

  as desired. □

Lemma 4. *For any formula $\eta \in \mathsf{Form}_{\mathcal{L}^+}$, the equivalence $\eta \leftrightarrow t(\eta)$ is a theorem of* H(**DCTSDAL**$^=$).

Proof. By induction on $n = \mathtt{w}(\eta)$, i.e., the number of occurrences of dynamic operators in η. When $n = 0$ as basis, the desired statement holds by Lemma 2.

For inductive step, suppose that, for any formula $\eta \in \mathsf{Form}_{\mathcal{L}^+}$ with $\mathtt{w}(\eta) \leqslant n$, we have established $\vdash^+ \eta \leftrightarrow t(\eta)$. We show that $\vdash^+ \eta \leftrightarrow t(\eta)$ for all formulas η such that $\mathsf{w}(\eta) \leqslant n+1$. This is shown by induction on the complexity $\mathsf{c}(\eta)$ (this is the second induction or subinduction). In what follows, we only deal with the cases where η is of the form $[\mathsf{Com}_{(x,y)}\varphi]P(x_1,\ldots,x_n)$, $[\mathsf{Com}_{(x,y)}\varphi]\Box\psi$, $[\mathsf{Com}_{(x,y)}\varphi]\mathsf{O}_{(x_1,x_2,x_3)}\psi$, or $[\mathsf{Com}_{(x,y)}\varphi][\mathsf{Prom}_{(z,u)}\chi]\psi$.

- Let η be of the form $[\mathsf{Com}_{(x,y)}\varphi]P(x_1,\ldots,x_n)$. By the definition of t, we show that $\vdash^+ [\mathsf{Com}_{(x,y)}\varphi]P(x_1,\ldots,x_n) \leftrightarrow (\top_{\{x,y\}\cup\mathsf{FV}(\varphi)} \to P(x_1,\ldots,x_n))$. But, this is the axiom (CPr) of H(**DCTSDAL**$^=$) (see Table 2).
- Let η be of the form $[\mathsf{Com}_{(x,y)}\varphi]\Box\psi$. We show that $\vdash^+ [\mathsf{Com}_{(x,y)}\varphi]\Box\psi \leftrightarrow t([\mathsf{Com}_{(x,y)}\varphi]\Box\psi)$. By definition, we show that $\vdash^+ [\mathsf{Com}_{(x,y)}\varphi]\Box\psi \leftrightarrow \Box t([\mathsf{Com}_{(x,y)}\varphi]\psi)$. By the axiom (C$\Box$), it suffices to prove $\vdash^+ \Box[\mathsf{Com}_{(x,y)}\varphi]\psi \leftrightarrow \Box t([\mathsf{Com}_{(x,y)}\varphi]\psi)$. By induction hypothesis for the second induction, we obtain $\vdash^+ [\mathsf{Com}_{(x,y)}\varphi]\psi \leftrightarrow t([\mathsf{Com}_{(x,y)}\varphi]\psi)$. It follows from Lemma 3 that $\mathsf{FV}([\mathsf{Com}_{(x,y)}\varphi]\psi) = \mathsf{FV}(t([\mathsf{Com}_{(x,y)}\varphi]\psi))$ hence $\vdash^+ \Box[\mathsf{Com}_{(x,y)}\varphi]\psi \leftrightarrow \Box t([\mathsf{Com}_{(x,y)}\varphi]\psi)$ by (RE$\Box$) of Proposition 2.
- Let η be of the form $[\mathsf{Com}_{(x,y)}\varphi]\mathsf{O}_{(x_1,x_2,x_3)}\psi$. We have to show that

$$\vdash^+ [\mathsf{Com}_{(x,y)}\varphi]\mathsf{O}_{(x_1,x_2,x_3)}\psi \leftrightarrow t([\mathsf{Com}_{(x,y)}\varphi]\mathsf{O}_{(x_1,x_2,x_3)}\psi).$$

By the definition of t and the axiom (CO) of Table 2, it suffices to establish the following two equivalences in H(**DCTSDAL**)$^=$:

$$\begin{aligned}&((x_1=y)\wedge(x_2=x)\wedge(x_3=x)\to\mathsf{O}_{(x_1,x_2,x_3)}(\varphi\to[\mathsf{Com}_{(x,y)}\varphi]\psi))\\ \leftrightarrow\ &((x_1=y)\wedge(x_2=x)\wedge(x_3=x)\to\mathsf{O}_{(x_1,x_2,x_3)}(t(\varphi)\to t([\mathsf{Com}_{(x,y)}\varphi]\psi))),\\ &(\neg((x_1=y)\wedge(x_2=x)\wedge(x_3=x))\to\mathsf{O}_{(x_1,x_2,x_3)}[\mathsf{Com}_{(x,y)}\varphi]\psi)\\ \leftrightarrow\ &(\neg((x_1=y)\wedge(x_2=x)\wedge(x_3=x))\to\mathsf{O}_{(x_1,x_2,x_3)}t([\mathsf{Com}_{(x,y)}\varphi]\psi)).\end{aligned}$$

First, we show the first equivalence. By induction hypothesis for the second induction, we obtain $\vdash^+ \varphi \leftrightarrow t(\varphi)$ and $\vdash^+ [\mathsf{Com}_{(x,y)}\varphi]\psi \leftrightarrow t([\mathsf{Com}_{(x,y)}\varphi]\psi)$. Then, the following implication hold:

$$\vdash^+ (x_1=y)\wedge(x_2=x)\wedge(x_3=x)\to((\varphi\to[\mathsf{Com}_{(x,y)}\varphi]\psi)\leftrightarrow(t(\varphi)\to t([\mathsf{Com}_{(x,y)}\varphi]\psi))).$$

Because $(\varphi \to [\mathsf{Com}_{(x,y)}\varphi]\psi)$ and $(t(\varphi) \to t([\mathsf{Com}_{(x,y)}\varphi]\psi))$ have the same set of free variables by Lemma 3, it follows from (CREO) of Proposition 2 that:

$$\begin{aligned}\vdash^+\ &\mathsf{O}_{(x_1,x_2,x_3)}((x_1=y)\wedge(x_2=x)\wedge(x_3=x))\\ &\to(\mathsf{O}_{(x_1,x_2,x_3)}(\varphi\to[\mathsf{Com}_{(x,y)}\varphi]\psi)\leftrightarrow\mathsf{O}_{(x_1,x_2,x_3)}(t(\varphi)\to t([\mathsf{Com}_{(x,y)}\varphi]\psi))),\end{aligned}$$

By (O =) from Proposition 2, we obtain:

$$\begin{aligned}\vdash^+\ &(x_1=y)\wedge(x_2=x)\wedge(x_3=x)\\ &\to(\mathsf{O}_{(x_1,x_2,x_3)}(\varphi\to[\mathsf{Com}_{(x,y)}\varphi]\psi)\leftrightarrow\mathsf{O}_{(x_1,x_2,x_3)}(t(\varphi)\to t([\mathsf{Com}_{(x,y)}\varphi]\psi))).\end{aligned}$$

This implies our desired first equivalence.
We move on to the second equivalence. It suffices to establish the following:

$$\begin{aligned}\vdash^+\ &\neg((x_1=y)\wedge(x_2=x)\wedge(x_3=x))\\ &\to(\mathsf{O}_{(x_1,x_2,x_3)}[\mathsf{Com}_{(x,y)}\varphi]\psi\leftrightarrow\mathsf{O}_{(x_1,x_2,x_3)}t([\mathsf{Com}_{(x,y)}\varphi]\psi)).\end{aligned}$$

By ($\mathsf{O} \neq$) from Proposition 2, it suffices to show the following:

$$\vdash^{+} \mathsf{O}_{(x_1,x_2,x_3)}\neg((x_1 = y) \land (x_2 = x) \land (x_3 = x)) \\ \to (\mathsf{O}_{(x_1,x_2,x_3)}[\mathsf{Com}_{(x,y)}\varphi]\psi \leftrightarrow \mathsf{O}_{(x_1,x_2,x_3)}t([\mathsf{Com}_{(x,y)}\varphi]\psi)).$$

But, this can be established similarly to the first equivalence.

– Let η be of the form $[\mathsf{Com}_{(x,y)}\varphi][\mathsf{Prom}_{(z,u)}\chi]\psi$. We show that

$$\vdash^{+} [\mathsf{Com}_{(x,y)}\varphi][\mathsf{Prom}_{(z,u)}\chi]\psi \leftrightarrow t([\mathsf{Com}_{(x,y)}\varphi]t([\mathsf{Prom}_{(z,u)}\chi]\psi)).$$

By induction hypothesis for the second induction, we have

$$\vdash^{+} [\mathsf{Prom}_{(z,u)}\chi]\psi \leftrightarrow t([\mathsf{Prom}_{(z,u)}\chi]\psi).$$

It follows that

$$\vdash^{+} [\mathsf{Com}_{(x,y)}\varphi][\mathsf{Prom}_{(z,u)}\chi]\psi \leftrightarrow [\mathsf{Com}_{(x,y)}\varphi]t([\mathsf{Prom}_{(z,u)}\chi]\psi),$$

because $[\mathsf{Com}_{(x,y)}\psi]$ is a normal modal operator (as in Remark 3). For our desired goal, it suffices to establish

$$\vdash^{+} [\mathsf{Com}_{(x,y)}\varphi]t([\mathsf{Prom}_{(z,u)}\chi]\psi) \leftrightarrow t([\mathsf{Com}_{(x,y)}\varphi]t([\mathsf{Prom}_{(z,u)}\chi]\psi)).$$

But, this is obtained from induction hypothesis for the first induction (on weight n), because $\mathsf{w}([\mathsf{Com}_{(x,y)}\varphi]t([\mathsf{Prom}_{(z,u)}\chi]\psi)) < \mathsf{w}([\mathsf{Com}_{(x,y)}\varphi][\mathsf{Prom}_{(z,u)}\chi]\psi) \leqslant n + 1$ by Lemma 3 hence $\mathsf{w}([\mathsf{Com}_{(x,y)}\varphi]t([\mathsf{Prom}_{(z,u)}\chi]\psi)) \leqslant n$. This finishes establishing the desired equivalence. □

Theorem 4. *Suppose that* H(**CTSDAL**$^{=}$) *is semantically complete, i.e., for every formula* $\varphi \in \mathsf{Form}_{\mathcal{L}}$, *if* φ *is valid then* φ *is a theorem of* H(**CTSDAL**$^{=}$). *Then,* H(**DCTSDAL**$^{=}$) *is also semantically complete.*

Proof. Suppose that H(**CTSDAL**$^{=}$) is semantically complete and fix any formula $\varphi \in \mathsf{Form}_{\mathcal{L}^{+}}$ such that φ is valid. By Lemma 4 and Theorem 3, $\varphi \leftrightarrow t(\varphi)$ is valid. Thus, $t(\varphi) \in \mathsf{Form}_{\mathcal{L}}$ is valid. By our initial supposition, $t(\varphi)$ is a theorem of H(**CTSDAL**$^{=}$). Since H(**DCTSDAL**$^{=}$) is an axiomatic expansion of H(**CTSDAL**$^{=}$), $t(\varphi)$ is also a theorem of H(**DCTSDAL**$^{=}$). By Lemma 4, we conclude that φ is a theorem of H(**DCTSDAL**$^{=}$), as required. □

5 Conclusion

We have established that Hilbert system H(**DCTSDAL**$^{=}$) is semantically complete relative to H(**CTSDAL**$^{=}$) (see Theorem 4). The language of **DCTSDAL**$^{=}$ enables us to state various general remarks concerning how acts of commanding and promising work as we have seen in Sect. 3. As is clear, however, there are many remaining tasks, questions and possibilities for further research.

First of all, as we have stated in Remark 2, the semantic completeness of H(**CTSDAL**$^{=}$) is still open. In this stage, what we have obtained so far is the semantic completeness of the fragment of H(**CTSDAL**$^{=}$) without the equality symbol as in [16].

Second, we may expand classes of formulas as given in Theorem 2 by keeping the validity of CGO and PGO principles. For example, the formula

$$[\mathsf{Com}_{(x,y)}\mathsf{O}_{(x,y,x)}(P(z) \to P(z))]\mathsf{O}_{(y,x,x)}\mathsf{O}_{(x,y,x)}(P(z) \to P(z))$$

is easily seen to be valid, but such formula is not covered by our syntactic conditions given in Theorem 2. But, we could not have CGO and PGO principles for the set of all formulas in the sense of Definition 5. This is because the following formula

$$[\mathsf{Com}_{(x,y)}\neg\mathsf{O}_{(y,x,x)}\neg P(y)]\mathsf{O}_{(y,x,x)}\neg\mathsf{O}_{(y,x,x)}\neg P(y)$$

is not valid, which is similar to the result that a Moore sentence is not successful [11, p.73, Example 4.5 and Sect. 4.7].

The remaining direction of further research is concerned with syntactic expansions of our language in this paper. As we only have variables as terms in our syntax for $\mathbf{DCTSDAL}^{=}$, one of the obvious remaining task is to introduce individual constants and function symbols into the language. We may also introduce epistemic modality, which enables us to capture effects of acts of requesting and asserting as the development of dynamic modal propositional logics of acts of requesting and asserting in [14,15] suggest. A slightly different possibility is that of introducing model updating operations different from the link-cutting utilized in $\mathbf{DCTSDAL}^{=}$. For example, see discussions of updates in [2,3,5].

Acknowledgements. We would like to thank the reviewers for their helpful comments and the participants of TLLM 2022 for their interesting questions and discussions. The work of both authors was partially supported by JSPS KAKENHI Grant Number 22H00597. The work of the first author was also partially supported by JSPS KAKENHI Grant Number 19K12113.

References

1. van Benthem, J.: Modal logic for open minds. CSLI Publications (2010)
2. van Benthem, J., Grossi, D., Liu, F.: Priority structures in deontic logic. Theoria **80**, 116–152 (2014)
3. van Benthem, J., Liu, F.: Dynamic logic of preference upgrade. J. Appl. Non-Classical Logics **17**(2), 157–182 (2007)
4. Fitting, M., Thalmann, L., Voronsky, A.: Term-modal logics. Stud. Logica. **69**, 133–169 (2001)
5. Hatano, R., Sano, K.: Recapturing dynamic logic of relation changers via bounded morphisms. Stud. Logica. **109**(1), 95–124 (2020). https://doi.org/10.1007/s11225-020-09902-5
6. Holliday, W.H., Icard, T.: Moorean phenomena in epistemic logic. In: Beklemishev, L.V.G., Shehtman, V. (eds.) Advances in Modal Logic, vol. 8, pp. 178–199. College Publications (2010)
7. Sawasaki, T., Sano, K., Yamada, T.: Term-sequence-modal logics. In: Blackburn, P., Lorini, E., Guo, M. (eds.) LORI 2019. LNCS, vol. 11813, pp. 244–258. Springer, Heidelberg (2019). https://doi.org/10.1007/978-3-662-60292-8_18
8. Sawasaki, T., Sano, K.: Frame definability, canonicity and cut elimination in common sense modal predicate logics. J. Log. Comput. **31**(8), 1933–1958 (2020). https://doi.org/10.1093/logcom/exaa067

9. Seligman, J.: Common sense modal predicate logic, draft, pp. 1–25 (2016)
10. Thalmann, L.: Term-modal logic and quantifier-free dynamic assignment logic, Ph. D. thesis, Uppsala University (2000)
11. van Ditmarsch, H., van der Hoek, W., Kooi, B.: Dynamic Epistemic Logic. Springer, Dordrecht(2007). https://doi.org/10.1007/978-1-4020-5839-4
12. Wang, Y., Cao, Q.: On axiomatizations of public announcement logic. Synthese **190**, 103–134 (2013)
13. Yamada, T.: Acts of promising in dynamified deontic logic. In: Satoh, K., Inokuchi, A., Nagao, K., Kawamura, T. (eds.) JSAI 2007. LNCS (LNAI), vol. 4914, pp. 95–108. Springer, Heidelberg (2008). https://doi.org/10.1007/978-3-540-78197-4_11
14. Yamada, T.: Acts of requesting in dynamic logic of knowledge and obligation. Eur. J. Anal. Phil. **7**(2), 59–82 (2011)
15. Yamada, T.: Assertions and commitments. Phil. Forum **47**(3–4), 475–493 (2016)
16. Yamada, T.: Completeness of CTSDAL. Presented in SOCREAL 2022 (6th International Workshop on Philosophy and Logic of Social Reality) (2022)

Incompleteness and (not-)at-issue Updates in Mandarin Chinese

Yenan Sun(✉)

The Chinese University of Hong Kong, Shatin, Hong Kong
yenansun@cuhk.edu.hk

Abstract. This paper re-examines the ongoing debate on whether overt aspect marking is required for expressing episodic readings in Mandarin Chinese [22,28,30,47,49,50,52,53]. I establish the generalization that overt aspect marking can be optional when the information concerning the occurrence of the event is NOT at-issue in the discourse, which is supported by both the new data on clause-embedding predicates and the existing observations in [37,45,46,50]. In addition, I show that zero-marked sentences in Mandarin are imperfective sentences which do not entail but are compatible with episodic interpretations. A formal pragmatic account is proposed to capture the discourse-sensitivity of such a requirement, making use of the unidimensional incremental dynamic theory of meaning in [2,3] and the Gricean theory of implicatures [5,19,27,42].

Keywords: Episodic sentence · Aspect · Incompleteness · At-issueness · Dynamic semantics · Gricean maxims

1 Introduction

Mandarin Chinese, as a morphologically tenseless language, relies on various mechanisms including lexical aspect, viewpoint aspect markers, temporal adverbials to express the temporal information of a sentence [37,45,46]. In particular, it has been argued that viewpoint aspect markers are generally required for (root) episodic sentences, namely those expressing the (partial) realization of a specific event within a contextually-relevant interval (i.e. Topic Time) [22,24,28,30,47,49,50,52,53]. A sentence involving a zero-marked eventive predicate such as (1) can only express a habitual reading (with appropriate frequency-indicating phrases) but not episodic readings as in (2), even with the presence of frame adverbials such as *zuotian* 'yesterday' in the latter.

(1) Lisi jingchang jian Mali.
Lisi often meet Mary
'Lisi often meets Mary'

(2) a. ??zuotian Lisi jian Mali.
yesterday Lisi meet Mary
Int: 'Yesterday Lisi met Mary' (Event-completion reading)

D. Deng et al. (Eds.): TLLM 2022, LNCS 13524, pp. 136–155, 2023.
https://doi.org/10.1007/978-3-031-25894-7_7

b. ??xianzai Lisi jian Mali.
now Lisi meet Mary
Int: 'Now Lisi is meeting Mary' (Event-in-progress reading)

In order to obtain episodic readings, overt (viewpoint) aspect markers such as the perfective *-le* or the progressive *zai* should be added, as in (3).[1] Otherwise, sentences such as those in (2) sound "incomplete".

(3) a. zuotian Lisi jian-le Mali.
yesterday Lisi meet-PERF Mary
'Yesterday Lisi met Mary'
b. xianzai Lisi zai jian Mali.
now Lisi PROG meet Mary
'Now Lisi is meeting Mary'

One prevailing explanation for the degradedness of the sentences in (2) (*incompleteness*, henceforth[2]) is that those sentences without overt aspect markers are syntactically or (/and) semantically flawed for the same reason that a bare sentence such as "Lisi meet Mary (yesterday)" is unacceptable in English. From the syntactic perspective, a bare sentence fails to project a certain size of structure (e.g. TP/IP, CP) that is required for being independent utterances [49,50]. From the semantic perspective, a sentence radical only describes a property of events and cannot denote the proposition that the event has occurred within the topic time without temporal operators such as aspect and tense [24,28,47].

However, what is tricky about Mandarin is that we do find many cases that are exempt from incompleteness.[3] It has been reported that a zero-marked sentence, when containing focus, can obtain episodic interpretations (4) [48,50].

(4) gangcai Lisi (shi) jian MALI.
just.now Lisi be meet Mary
'Just now it is [Mary]$_F$ that John {met/was meeting}.'[4]

[1] When overt aspect markers are present, frame adverbials are largely optional. Though see [23,35,50] for discussion on certain sentences with the perfective *-le* and a dummy object, which could still sound incomplete without frame adverbials.

[2] I use "incompleteness" as a descriptive label and will not discuss other kinds of "incomplete" sentences [22,35,50,52] in this paper. The reason is that without a careful examination and comparison of those phenomena, one cannot conclude that all sentences that sound incomplete intuitively should receive the same analysis.

[3] In fact, some authors [37,45] claim that viewpoint aspect markers are generally optional for episodic sentences though such a claim has been argued to be empirically incorrect due to the degradedness in (2), see [24,47,53].

[4] In particular, the meaning of (4) is underspecified between an event-completion reading and an event-in-progress reading, which accords with [45]'s observation about Mandarin zero-marked sentences. According to [37,46], which reading is more salient is subject to a variety of factors such as context, world knowledge, the length of the Topic Time, and the lexical aspect of the predicate. For the examples of zero-marked sentences in the rest of the paper, I will just list one available (and usually salient) reading but it does not mean the listed one is the only possible reading.

[7,45,46,54] also point out in certain narratives such as (5), zero-marked sentences can also obtain episodic interpretations.

(5) jintian zaoshang Lisi paobu, chi zaofan, ranhou qu xuexiao.
today morning Lisi ran eat breakfast then go school
'This morning, Lisi ran, ate breakfast, and then went school.'

The existing syntactic-semantic approaches are problematic in that they either ignore the data like (4)–(5) and claim that incompleteness exists across the board [28,47], or they have to assume that adding focus is another way of satisfying a certain syntactic/semantic requirement satisfied by the aspect markers in (3). A representative of the latter by [50] (see also [22]) proposes that a sentence can either be "anchored" by temporal operators (in the sense of [12]) or by focus operators. One concern for this kind of approach is that it is cross-linguistically uncommon for focus to have the same syntactic/semantic functions as aspect markers, leaving this binary condition of anchoring unmotivated.[5] Moreover, the binary condition is still too strong in that many narration examples given in [45,46,54], including (5), have neutral intonation and can be uttered in out-of-the-blue contexts; therefore their acceptability cannot be straightforwardly attributed to the presence of focus.

The goal of this paper is to argue that incompleteness in Mandarin can be better captured by a pragmatic approach. The motivations are twofold. Firstly, I establish the novel generalization that incompleteness is sensitive to what information is at-issue in the discourse, based on both the novel piece of data on clause-embedding predicates and the existing observations such as (4)–(5) [37,45,46,54]. Secondly, there is reason to believe that those apparently bare sentences are in fact syntactically and semantically well-formed – they can express habitual readings and are just constrained in terms of expressing episodic readings. Based on those motivations, a formal dynamic pragmatic account is proposed to capture the discourse-sensitivity of incompleteness.

The paper is organized as follows. Sections 2 and 3 present two motivations of pursuing a pragmatic approach of incompleteness. Section 4 proposes a formal pragmatic analysis, making use of the unidimensional incremental dynamic theory of meaning in [2,3] and the Gricean theory of implicatures [5,19,27,42]. Section 5 concludes.

2 Incompleteness Correlates with What is at-Issue

In this section, I present new data concerning clause-embedding predicates in Mandarin to advance the novel generalization in (6).

[5] [50] attributes the cross-linguistic difference between Mandarin and English in the availability of focus anchoring to the parametric variation in the (non)-existence of empty particles; but they do not discuss how focus anchoring contributes to an episodic reading semantically.

(6) Incompleteness arises ONLY when the inference concerning the occurrence of the event (described by the zero-marked predicate) is the main point of the utterance, or is at-issue.

Then I show that this generalization can also capture the existing observations that zero-marked sentences with focus or those within certain narratives do not suffer from incompleteness [45,46,50].

How do one tell whether an inference of a certain utterance is at-issue or not? One useful diagnostics is that at-issue inferences can address the Question Under Discussion in the discourse [29,41,44,51]. In particular, we can form a question-answer pair to see whether the utterance can be a felicitous response to a question to which that inference is relevant.[6] Consider an English sentence containing an appositive such as (7). While one can infer both (7-a) and (7-b) from such an utterance, the discourse statuses of those two inferences are not equivalent – only the former is considered to be at-issue since this utterance is a felicitous response to a question to which the inference (7-a) is relevant, but not to a question to which the inference (7-b) is relevant, c.f. (8), (9).

(7) Mary, who just defended her thesis, got a job recently.
 a. ⇝ Mary got a job recently.
 b. ⇝ Mary just defended her thesis.

(8) Q: Did anyone get a job?
 A: Mary, who just defended her thesis, got a job recently.

(9) Q: Did anyone defend their thesis?
 A: #Mary, who just defended her thesis, got a job recently.

We will mainly rely on this diagnostic in the rest of the section.

2.1 New Data Concerning Clause-Embedding Predicates

When an eventive verb embeds a (complete[7]) clausal complement, overt aspect marking on that matrix verb can be optional for expressing episodic readings, as in (10). Note that *shuo* 'say' in Mandarin is a regular eventive verb since if we replace the clausal complement with a nominal object as in (11), the incompleteness effect comes back.[8]

(10) Lisi shuo-(le) [jintian hen re]
 Lisi say-PERF today very hot
 'Lisi said that it is hot today'

[6] For the purpose of this paper, it is sufficient to adopt the following simplified definition of relevance: a proposition p is relevant to a question Q if p contextually entails a (partial) answer of Q.

[7] I call the embedded stative clause as "complete" because stative sentences do not need aspect markers and they can always stand as independent utterances.

[8] While there is no space to present more data in this paper, note that the contrast like (10) and (11) is prevalent for a large set of clause-embedding eventive verbs in Mandarin including *gaosu* 'tell', *tingshuo* 'hear', *xuanbu* 'announce'.

(11) zuotian Lisi shuo-??(le) zhe-ju hua.
yesterday Lisi say-PERF this-CL sentence
Int: 'Lisi said this sentence yesterday'

I propose that overt aspect marking is optional there because (10) can present either the embedded or the matrix content as the at-issue proposal (i.e. the main point), which is common for biclasual constructions cross-linguistically [40,43].

When we make the Question Under Discussion (QUD) explicitly concern the embedded content of (10), *-le* is indeed optional on *shuo*, as in (12). In this case, the matrix predication does not need to be the main point of the utterance but can serve a secondary, evidential-like function, namely specifying the source of the information expressed by the embedded sentence.

(12) Q: jintian tianqi zenmeyang?
today weather how
'What is the weather like today?'
A: ✓(10) without *-le*; ✓(10) with *-le*

By contrast, when the QUD is explicitly made to concern the realization of the matrix event as in (13), it is pragmatically odd to answer the question with the aspectually zero-marked version of (10).

(13) Q: wo huibao tianqi shi ni weishenme zheme bu naifan?
I report weather when you why so not patient
'Why were you so impatient when I was reporting the weather?'
A: #(10) without *-le*; ✓(10) with *-le*

The contrast between (12) and (13) shows that when *-le* is omitted in (10), the inference concerning the occurrence of the saying event is not at-issue. In other words, embedding a clausal complement makes it possible to put forth the content of that clausal complement as the at-issue proposal and in this case the matrix proposition serves as a secondary point like an evidential phrase. For this reason, overt aspect markers can be optional on the matrix eventive verbs.

Turning to the monoclausal use of the verb *shuo* 'say' in (11), or other monclausal sentences uttered with neutral intonation in general (e.g. (2)), incompleteness arises because in those cases the occurrence of the matrix event is by default at-issue in an out-of-the-blue context. This intuition is reflected in the assumption of the literature that the default QUD in such a context is 'What happened?' or 'What's new?' [1,14,31,41]. The reason why many authors treat incompleteness as a context-free grammaticality requirement is because they happen to focus on monoclausal sentences uttered in out-of-the-blue contexts such as (11) and (2).

2.2 Extending to the Existing Observations

This section shows that the proposed generalization in (6) can further capture the existing observations that narratives and focus can salvage incompleteness.

Within Narratives. When a zero-marked eventive sentence is uttered with other matrix sentences to form a sequence (i.e. a narrative), overt aspect marking can be optional for episodic interpretations. In (14), the zero-marked sentence is followed by a complete sentence (e.g. a zero-marked stative sentence or an aspectually marked eventive one), and the entire utterance can be acceptable.

(14) zuotian Lisi jian Mali, ta hen kaixin.
yesterday Lisi meet Mary he very happy
'Yesterday Lisi met Mary, he was happy.'

[7] provides many examples from corpora to support this tendency (see also [46, 54]). One example reproduced in (15) is from Lu Xun's novel *Yao* "Medicine": all the events described in the narrative indeed occurred, and only the last verb *miman* "fill" is overtly marked with the perfective.

(15) Hua Laoshuan turan zuoqi shen, cazhao huochai, dian.shang
Hua Laoshuan suddenly sit.up body strike match light.on
bian.shen youla de dengzhan, chaguan de liang-jian wuzi.li
whole.body grease DE oil-lamp tea.house DE two-CL room.LOC
bian mimian-le qing.bai de guang.
then fill-PERF green-white DE light
'Hua Laoshuan suddenly sat up, (and) striking a match, lit the completely grease-covered oil lamp. The two rooms in the teahouse then were filled with a greenish-white light'

I argue that zero-marked eventive sentences are not degraded in cases like (14) and (15) because it is possible to have the other sentence(s) to contribute the at-issue update of the discourse while making the realization of the event described by the zero-marked predicate a secondary point of the utterance. Take (14) for instance, that its first clause is not at-issue can be confirmed by the infelicity of uttering (14) as a response to the question 'Did Lisi met anyone?':

(16) Q: Lisi you jian renhe ren ma?
Lisi PERF meet any person YNQ
'Did Lisi meet anyone?'
A: #zuotian Lisi jian Mali, ta hen kaixin.
'Yesterday Lisi met Mary, he was happy.'

In contrast, there is no problem of uttering (14) as a response to the question 'How's Lisi's mood?', as in (17).

(17) Q: Lisi xinqing zenmeyang?
Lisi mood how
'How is Lisi's mood?'
A: zuotian Lisi jian Mali, ta hen kaixin.
'Yesterday Lisi met Mary, he was happy.'

Another typical kind of narrative in which incompleteness disappears is a run-on sentence like (18), in which all the predicates are zero-marked but it conveys that a sequence of events occurred one after one this morning.

(18) jintian zaoshang Lisi paobu, chi zaofan, ranhou qu xuexiao.
today morning Lisi ran eat breakfast then go school
'This morning, Lisi ran, ate breakfast, and then went school'

This kind of utterance naturally occurs in a context in which it is taken for granted that a sequence of events has occurred during the topic time (e.g. when reporting one's daily routine), and the main point of uttering (18) is to inform about the content of each event and their order.[9] It is indeed quite odd to use (18) to answer a neutral polar question such as (19), which concerns whether each relevant event is actualized or not.

(19) Q: jintian Lisi you mei you {paobu / chi zaofan / qu xuexiao}?
today Lisi have not have run / eat break / go school
'Did Lisi {run / eat breakfast / go to school} today?'
A: #jintian zaoshang Lisi paobu, chi zaofan, ranhou qu xuexiao.

In words, zero-marked episodic sentences within certain narratives are not degraded because it is possible to have the occurrence of the event expressed by the zero-marked predicate as not-at-issue information in those cases.

The Presence of Focus. Sentences with focus are often exempt from incompleteness (see also [22,50]), as in (20).

(20) zuotian Lisi (shi) jian MALI.
yesterday Lisi be meet Mary
'It is $[\text{Mary}]_F$ that Lisi met yesterday'

I propose that incompleteness disappears in those cases because with the presence of focus, the occurrence of some event is in fact presupposed in the context, and presupposed contents are typically not at-issue. Indeed, the sentences in (20) are naturally uttered in contexts in which it is already in the common ground that some event has occurred. In (21), it is already known that Lisi met someone yesterday, and the main point of B's utterance concerns who Lisi met.

(21) A: zuotian Lisi jian-le yi-ge ren.
yesterday Lisi meet-PERF one-CL person
'Yesterday Lisi meet a person'

[9] One reviewer raises a similar example in which the utterance contains multiple sentences and none of the verbs are aspectually marked. In particular, that example serves as an answer to a *why*-question, which strongly presupposes the occurrence of some event as the cause and the at-issue update concerns the event content instead of its occurrence. For this reason, I do not think those examples threaten the current generalization.

B: shide. zuotian ta jian MALI.
'Right. It is [Mary]$_F$ he met yesterday'

The presupposed status of the event occurrence inference in those examples can be further confirmed by their ability to project over entailment-canceling operators, as illustrated in (22).

(22) zuotian Lisi jian MALI ma?
yesterday Lisi meet Mary YNQ
'Is it [Mary]$_F$ that Lisi met yesterday?'
⇝ Lisi met someone yesterday.

In words, focused sentences are exempt from incompleteness because in those cases the information that some event has occurred within the topic time is not at-issue (i.e. presupposed).

2.3 Interim Summary

This section shows that what is shared by the cases in which incompleteness disappears is that the information concerning the occurrence of the event (described by the zero-marked predicate) is NOT at-issue: embedding a clausal complement makes it possible for the matrix predication to serve an evidential-like function, which is a secondary point of the utterance; creating a narrative can shift the main point of the utterance to be some other inferences; and adding focus renders such information presupposed.

3 Zero-Marked Sentences are Imperfective Sentences

Besides the discourse-sensitivity of incompleteness illustrated in Sect. 2, this section provides another reason why a pragmatic account might be appealing – zero-marked sentences in Mandarin are in fact syntactically and semantically well-formed (under certain readings).

3.1 Imperfective Uses of Mandarin Zero-Marked Sentences

Zero-marked sentences in Mandarin are typically used to convey habitual readings as in (1) (repeated as (23)), futurate readings such as (24), and continuous readings (for stative predicates) as in (25).

(23) Lisi jingchang jian Mali.
Lisi often meet Mary
'Lisi often meets Mary'

(24) mingtian Lisi jian Mali.
tomorrow Lisi meet Mary
'Lisi meets Mary tomorrow'

(25) Lisi dong fayu.
Lisi know French
'Lisi knows French.'

Those readings are usually expressed by the imperfective form across languages, for example, the simple form in English in (26), or the imperfective form in languages such as Russian and Spanish [17,21].

(26) a. Lisi often meets Mary.
b. Lisi meets Mary tomorrow.
c. Lisi lives in Suzhou.

Together with the fact that Mandarin zero-marked sentences are indeed acceptable for episodic readings in a lot of cases (as shown in Sect. 3), one hypothesis is that zero-marked sentences are grammatical in the first place and they are just the counterparts of imperfective forms in other languages. They cannot freely express episodic readings just like how imperfective forms in many languages do not typically express episodic readings but can do so in a restricted way [4,9,17,21]. For instance, the imperfective form in French is not typically used for expressing event-completion readings, but it can express perfective-like readings in a narrative such as (27), which is quite similar to how Mandarin zero-marked sentences can express episodic readings in run-on sentences.

(27) A huit heures, les voleurs entraient dans la banque, ils
At eight hours, the robbers entered(Impf) in the bank, they
discutaient avec un employé, puis se dirigeaient vers
discussed(Impf) with an employee, then Refl directed(Impf) towards
le guichet principal.
the window main
'At eight, the robbers entered the bank, they discussed with a clerk, then they moved towards the main desk.' (Adapted from [26])

The imperfective form in Spanish also cannot freely express event-in-progress readings but [17] claims that it can do so when all discourse participants have perceptual access to the ongoing event. The mutual perceptual access in the context makes it easy to accommodate the occurrence of the relevant event, and this condition resembles how zero-marked sentences with focus can express episodic readings in Mandarin.

While the specific constraints of using imperfective forms to express episodic readings in those languages are not identical to that in Mandarin, they do overlap, which points to the possibility that zero-marked sentence in Mandarin are grammatical and meaningful imperfective forms which are just constrained in terms of conveying episodic readings.

3.2 The Literal Meaning of Zero-Marked Imperfective Forms

If a zero-marked sentence is an imperfective sentence, then what kind of meaning does it express? Following a standard intensional analysis of the imperfective aspect [4,9–11,15,16,33], I argue that it does not anchor the event described by the predicate to the evaluation world, but instead to the inertia continuations of the evaluation world (since the topic time), namely the possible worlds that are identical to the evaluation world w before the topic time and develop in ways most compatible with the regular course of the relevant affairs since the topic time. The lexical entry of a null imperfective morpheme in Mandarin, $\varnothing_{\text{IMPF}}$, is given in (28).

(28) $$[\![\varnothing_{\text{IMPF}}]\!] = \lambda P_{\langle s,vt \rangle} \lambda w \lambda i. \forall w' \in \textbf{INERT}(w, i) : \exists e[P(e, w') \wedge \tau(e, w') \supseteq i]$$

The modal base "**INERT**(w, i)" takes the evaluation world w and the topic time i and returns a set of inertia worlds. This modal component is crucial in that it captures the intensional character of the imperfective uses. For a habitual sentence such as 'Lisi meets Mary', it entails a regular occurrence of the event of Lisi meeting Mary not necessarily in the actual world but in an ideal world in which nothing interrupts the normal schedule of Lisi and Mary (and any other relevant affairs). For a futurate sentence like 'Lisi meets Mary tomorrow', it entails the occurrence of Lisi meeting Mary tomorrow in ideal worlds in which the plan made at or before the speech time is not interrupted. For stative sentences [15,16] also argue that the lexical entry in (28) can capture the ongoingness of a state during the topic time, except that the inertia modality is vacuous due to the different properties of states compared to events.

This paper does not intend to discuss the formal details of how (28) can capture the different imperfective uses, and the interested readers can refer to [9,11,15,16,33]. What is crucial to the current discussion is that, an imperfective form, even when the topic time is overtly restricted by the frame adverbials such as *gangcai* 'just now' or *xianzai* 'now', does not entail the occurrence of the event in the actual world. The sentence in (29) for instance, can at best express a futurate-like reading as in (30), and due to the modal nature of imperfective, it does not entail the occurrence of the meeting event in the actual world.

(29) gangcai Lisi jian Mali.
just.now Lisi meet Mary
lit: 'If things went inertially since just now, the event of Lisi meeting Mary occurred within just now'

(30) a. $[\![[\varnothing_{\text{IMPF}}\ [\text{Lisi meet Mary}]]]\!] = \lambda w \lambda i. \forall w' \in \textbf{INERT}(w, i) : \exists e[\textbf{meet}(e, l, m, w') \wedge \tau(e, w') \supseteq i]$

b. Via some covert tense/semantic rule:[10]
$[\![(29)]\!] = \lambda w. \exists i[i \subseteq \textbf{just.now} \wedge \forall w' \in \textbf{INERT}(w, i) : \exists e[\textbf{meet}(e, l, m, w') \wedge \tau(e, w') \supseteq i]]$

[10] The temporal interval variable i in (30-a), which represents the topic time, is existentially bound either by some covert tense operator as in [24,47] or via some semantic rule as in [34,37,46]. This particular choice of how to relate topic time to the speech time is orthogonal to the current investigation, and for this reason, I will directly present the result of this semantic process as in (30-b).

Still, this semantics is compatible with the scenarios in which the meeting event occurred just now – it is clearly possible that nothing interrupts the normal course of the relevant affairs in the actual world, and in this case the meeting event should have occurred within the topic time.

3.3 Interim Summary

This section showed that the literal meaning of a zero-marked sentence involves imperfective semantics and is truth-conditionally weaker than the episodic reading that the relevant event is instantiated within the topic time in the evaluation world, but is nevertheless compatible with it. Importantly, zero-marked sentences are not ungrammatical, but are just not typically used for expressing episodic readings, which is quite common cross-linguistically.

4 A Formal Pragmatic Account

This section proposes a formal pragmatic account of Mandarin incompleteness. The main idea is that a zero-marked sentence can obtain episodic readings because its imperfective semantics can be strengthened into such readings via some pragmatic principle (i.e. Gricean Maxims [5,19]). However, such enrichment can be blocked due to the competition between the zero-marked imperfective form and the better alternatives of expressing episodic readings. The blocking effect is obligatory when such information is directly relevant to the QUD, which results in the observed discourse-sensitivity of incompleteness.

Section 4.1 briefly introduces the basic setup for the formal dynamic implementation. Section 4.2 proposes that zero-marked sentences can express episodic readings at least in certain cases via pragmatic enrichment, while such pragmatic enrichment is not always available due to aspectual competition.

4.1 Formal Setup

Since we will need to formally distinguish between at-issue and not-at-issue information in the discourse, and keep track of the anaphoric information across at-issue and not-at-issue updates, I adopt a unidimensional incremental dynamic theory of meaning in [2,3] (which is based on [13,20,41]).

[3] propose to treat discourse contexts as sets containing the classic Stalnakerian Context Set and all their subsets, and use the designated propositional variable p^{cs} to store the current Context Set (CS). For an at-issue assertion, it puts forth a proposal, which is stored with a propositional discourse referent (*dref* henceforth), p, to update the CS by restricting possible future contexts to those that have non-empty intersections with p, namely $p^{cs} \cap p$. If accepted, the CS is updated by assigning a new value to the dref p^{cs} (':=' is used to indicate (re)assignment of values to variables):

(31) $p^{cs} := p^{cs} \cap p$

For not-at-issue updates, such as those contributed by appositive or evidential content, they do not put forth a proposal but directly eliminate the assignments that assign to p^{cs} at least one world in which the proposition is not true, and this process is called "an imposal" by [3]. This reflects the distinction made in [13] that not-at-issue updates happen automatically without regular negotiation associated with at-issue updates. For presuppositions, which are not at-issue and old information, they are treated as preconditions on the input CS.

The dynamic theory is implemented in an extension of Dynamic Predicate Logic [2,3,20]. Our models consist of domains of individuals $\mathcal{D}$, eventualities $\mathcal{E}$, temporal intervals $\mathcal{T}$, and possible worlds $\mathcal{W}$, and the interpretation function $\mathcal{I}$ that assigns a subset of $\mathcal{D}_n$ to any n-ary relation $\mathcal{R}$ relative to any world w, i.e., $\mathcal{I}_w(\mathcal{R}) \subseteq \mathcal{D}^n$. We have variables over individuals $(x, y, z...)$, eventualities $(e_1, e_2, ...)$, temporal intervals $(i_1, i_2, ...)$, and worlds $(w, w', ...)$ and propositions/sets of worlds $(p, p', p^{cs}, q, l, m, ...)$, and the usual inventory of non-logical constants: individual constants (**John**, . . .), properties (**person**, **yesterday**, ...), n-relations (**meet**, ...), etc.

Formulas are interpreted relative to a pair of assignments $\langle g, h\rangle$, i.e., they denote functions from an input assignment g to an output assignment h. New variables are introduced via random assignments as in (32). In particular, certain variables can be relativized to worlds (33):

(32) $[\![[p]]\!]^{\langle g,h\rangle} = 1$ iff for any variable p' s.t. $p' \neq p$: $g(p') = h(p')$

(33) $[\![[x_p]]\!]^{\langle g,h\rangle} = 1$ iff

a. for any variable x' s.t. $x' \neq x$: $g(x') = h(x')$, and

b. $\begin{cases} \mathbf{Dom}(h(x)) = h(p^{cs}) \text{ if } p \subseteq p^{cs} \text{ is the at-issue proposal} \\ \mathbf{Dom}(h(x)) = h(p) \text{ otherwise} \end{cases}$

Atomic formulas are defined if the preconditions are satisfied, and lexical relations are interpreted relative to propositions, as in (34):

(34) a. $[\![R_p(x_1, ..., x_n)]\!]^{\langle g,h\rangle}$ is defined iff for any $i \in \{1, .., n\}, h(p) \subseteq \mathbf{Dom}(h(x_i))$.

b. If defined, $[\![R_p(x_1, ..., x_n)]\!]^{\langle g,h\rangle} = 1$ iff $g = h$ and for all worlds $w \in h(p)$: $\langle h(x_1)(w), ..., h(x_n)(w)\rangle \in \mathcal{I}_w(R)$

For complex formula, (35) provides a partial list of the interpretation rules (adapted from [3]): dynamic conjunction in (35a); negation in (35b); the inertia modal base in the imperfective semantics which is relativized to both worlds and time intervals as in (35c).

(35) a. $[\![\phi \wedge \psi]\!]^{\langle g,h\rangle} = 1$ iff there exists an assignment k such that $[\![\phi]\!]^{\langle g,k\rangle} = 1$ and $[\![\psi]\!]^{\langle k,h\rangle} = 1$.

b. $[\![\mathbf{NOT}_p^{p'}(\phi)]\!]^{\langle g,h\rangle} = 1$ iff

(i) $[\![[p'] \wedge \phi]\!]^{\langle g,h\rangle} = 1$, and there is no h' s.t. $[\![[p'] \wedge \phi]\!]^{\langle g,h'\rangle} = 1$ and $h(p') \not\subseteq h'(p')$ and

(ii) $h(p) \cap h(p') = \emptyset$

c. $[\![\mathbf{INERT}_{p,i}^{p'}(\phi)]\!]^{\langle g,h\rangle} = 1$ iff

(i) $[\![[p'] \wedge \phi]\!]^{\langle g,h \rangle} = 1$, and there is no h' s.t. $[\![[p'] \wedge \phi]\!]^{\langle g,h' \rangle} = 1$ and $h(p') \not\subseteq h'(p')$ and
(ii) for all $w \in h(p)$, $\mathbf{INERT}(w, i) \supseteq h(p')$

4.2 Pragmatic Enrichment and Aspectual Competition

This section aims to explain why zero-marked sentences, which do not entail the event realization inference as shown in Sect. 3, can express episodic readings, but only when such an inference is not at-issue.

I start with the key piece of data (the evidential case) in (36). I follow Murray's analysis of evidentiality in [40] such that the matrix proposition is imposed on the CS (37a, 37c), while a modalized version of the embedded proposition is put forth as the at-issue proposal (37b). I leave out the modal part of the at-issue proposal for simplicity since it is irrelevant to our discussion.

(36) Lisi ∅_{IMPF} shuo [jintian hen re].
Lisi say today very hot
'Lisi said that it is hot today'

(37) Update the CS with (36) (in its evidential use):
a. Imposal: $[x] \wedge x = \mathbf{Lisi} \wedge [i_1] \wedge$
b. Proposal: $[q] \wedge q \subseteq p^{cs} \wedge$
$[i_2] \wedge i_2 \subseteq \mathbf{today} \wedge [s_1] \wedge \mathbf{hot}_q(s_1) \wedge \tau_q(s_1) \supseteq i_2 \wedge$
c. Imposal: $\mathbf{INERT}^{p'}_{p^{cs},i_1}([e_{1_{p'}}] \wedge \mathbf{say}_{p'}(e_1, x, q) \wedge \tau_{p'}(e_1) \supseteq i_1) \wedge$
d. $\rightsquigarrow_{Quantity_2}$ $[e_1] \wedge \mathbf{say}_{p^{cs}}(e_1, x, q) \wedge \tau_{p^{cs}}(e_1) \supseteq i_1$

Since the matrix verb *shuo* combines with a null imperfective, it does not entail the occurrence of the saying event in the actual world, but only says the event has occurred within those ideal inertia worlds in which the relevant affairs develop in a regular course. Crucially, I follow [6] in arguing that the default assumption, when no other information uttered by the speaker or in the context contracts it, is that the actual world indeed develops inertially so that the saying event has occurred within the topic time (i_1), as represented in (37-d).[11] I further identify the source of this reasoning as the Gricean Maxim of Quantity-2 'Say no more than you need' [19]. [5,25] elaborate it as 'What is stereotypical needs not be said'. (38) for instance, by default invites the implicature that John broke his own finger, since breaking one's own finger is the stereotypical, unmarked situation. I argue that the same kind of reasoning is at play for the zero-marked imperfective sentence in (36).

(38) John broke a finger.
$\rightsquigarrow_{Quantity_2}$ John broke his finger.

Indeed, we find the event realization inference expressed by the Mandarin zero-marked sentence can be defeasible, as long as it is compatible with the contextual

[11] [6] uses this reasoning to explain why accomplishments in St'át'imcets and Skwxwú7mesh can invite a default culmination inference, though defeasible.

information. (39) is such an example in which the zero-marked sentence is continued with a sentence that contradicts the event realization inference.

(39) gangcai Mali gen laoshi shuo [Yuehan da-le ren]. dan hai
just.now Mary to teacher say John beat-PERF person but yet
mei kaikou jiu bei Bi'er zuzhi le.
not open.mouth JIU BEI Bill stop LE
'Just now Mary was about to tell the teacher that [John beat a person]'. But Bill stopped her before she opened her mouth.'

The question is, if a zero-marked sentence can obtain episodic readings via this pragmatic enrichment, why such reading is not always available? In particular, why is a zero-marked sentence uttered out-of-the-blue (repeated as in (40)) considered to be incomplete?

(40) ??gangcai Lisi $\varnothing_{\text{IMPF}}$ jian Mali.
just.now Lisi meet Mary
Int: 'Just now Lisi met Mary'

Representing the update made by the utterance in (40) formally as in (41-a), we find the main difference between (36) and (40) lies in the whether the occurrence of the event within the topic time is part of the at-issue proposal or not.

(41) a. Proposal: $[p] \wedge p \subseteq p^{cs} \wedge [x] \wedge x = \textbf{Lisi} \wedge [y] \wedge x = \textbf{Mary} \wedge [i_1] \wedge i_1 \subseteq \textbf{just.now} \wedge \textbf{INERT}^{p'}_{p,i_1}([e_{1_{p'}}] \wedge \textbf{meet}_{p'}(e_1, x, y) \wedge \tau_{p'}(e_1) \supseteq i_1)$
b. (In principle possible:)
$\rightsquigarrow_{Quantity_2} [e_1] \wedge \textbf{meet}_p(e_1, x, y) \wedge \tau_p(e_1) \supseteq i_1$

I argue that (41-a) fails to obtain episodic readings via (41-b) because when the occurrence of the event is at-issue as in (41-a), the 'better' alternatives are mandatorily evoked, namely those with overt aspect markers. According to the Gricean Maxim of Quantity-1 'Say as much as you can (if relevant)' and the subsequent neo-Gricean theories, the speaker is expected to use the truth-conditionally stronger but no more complex alternatives (when relevant), and if not, there arises a Quantity-1 implicature that it is not the case that the stronger alternatives are true [19,27,42]. Now let us check if the alternatives marked by the overt perfective or progressive are both truth-conditionally stronger and no more complex than the imperfective form.

First, following a standard analysis of viewpoint aspects ([24,36,45,47,55]), the sentence involving the perfective *-le* entails the completion of the event within the topic time in the evaluation world (42); the sentence involving the progressive *zai* entails the partial realization of the eventuality in the evaluation world (43). The variations between the existing analyses are not our concern: what is uncontroversial is that the sentences involving the overt perfective or progressive markers must be truth-conditionally stronger than the sentences involving the zero-marked imperfective morpheme, since the latter does not entail even the partial realization of the event in the evaluation world (44).

(42) $[\![\text{just.now}[\text{-PERF}[\text{Lisi meet Mary}]]]\!] = \lambda w.\exists i[i \subseteq \textbf{just.now} \wedge \exists e[\textbf{meet}(e,l,m,w) \wedge \tau(e,w) \subseteq i]$

(43) $[\![\text{just.now}[\text{PROG}[\text{Lisi meet Mary}]]]\!] = \lambda w.\exists i[i \subseteq \textbf{just.now} \wedge \exists e'[\tau(e', w) \subseteq i \wedge \forall w' \in \textbf{INERT}(w,i) : \exists e[e' \sqsubseteq e \wedge \textbf{meet}(e,l,m,w') \wedge \tau(e,w') \supseteq i]]]$

(44) $[\![\text{just.now}[\varnothing_{\text{IMPF}}[\text{Lisi meet Mary}]]]\!] = \lambda w.\exists i[i \subseteq \textbf{just.now} \wedge \forall w' \in \textbf{INERT}(w,i) : \exists e[\textbf{meet}(e,l,m,w') \wedge \tau(e,w') \supseteq i]]$

Second, adopting the concept of structural complexity in [27], the alternatives in (42) and (43) are no more complex than (44) because they can be derived from the syntactic structure of the uttered sentence (44) by substituting the covert imperfective morpheme with overt aspectual morphemes. Note that since the complexity of alternatives based on their structure, the zero phonology of the imperfective morpheme does not factor into the complexity measure since (44) projects a structure of the same size as the overtly marked alternatives.

In words, not using those "better" alternatives when relevant potentially gives rise to an implicature that contradicts (41-b), as represented in (45)

(45) $\rightsquigarrow_{Quantity_1} \textbf{NOT}_p^{p'}([e_{1_{p'}}] \wedge \textbf{meet}_{p'}(e_1, x, y) \wedge \tau_{p'}(e_1) \subseteq i_1)$

I propose that this "problematic" Quantity-1 implicature is mandatorily triggered when those better alternatives are directly relevant to the Question Under Discussion (see similar assumptions in [8,25,32]), alternatively speaking, when the semantic contribution of the imperfective is part of the at-issue proposal. In the case of (40), since the information contributed by the imperfective is at-issue, and the Quantity-1 implicature is mandatory and blocks the potential pragmatic enrichment via Maxim of Quantity-2. In the case of uttering the biclausal construction such as (36), since the information contributed by the imperfective is imposed, the Quantity-1 implicature is optional, which makes the pragmatic enrichment possible. For this reason, zero-marked sentences can only express episodic readings in restricted cases, namely when the information concerning the event occurrence is not at-issue.

The rest of this section illustrates how the proposed analysis can extend to the cases in which incompleteness disappears, namely zero-marked sentences in narratives or those with focus.

I have shown in Sect. 2 that, in a narrative such as (46), the proposition denoted by the first sentence is not-at-issue while the second stative sentence constitutes the at-issue proposal. The discourse update contributed by the entire (46) can be formalized in (47). Since the information contributed by the imperfective is imposed as in (47a), the better alternatives marked by the overt aspect markers are not directly relevant to the QUD and the potential Quantity-1 implicature is optional. For this reason, pragmatic enrichment via the Maxim of Quantity-2 is not blocked.

(46) zuotian Lisi $\varnothing_{\text{IMPF}}$ jian Mali. ta hen kaixin.
yesterday Lisi meet Mary he very happy
'Yesterday Lisi met Mary. He was happy.'

(47) a. Imposal: $[x] \wedge x = \mathbf{Lisi} \wedge [y] \wedge x = \mathbf{Mary} \wedge [i_1] \wedge i_1 \subseteq \mathbf{just.now} \wedge$ $\mathbf{INERT}^{p'}_{p^{cs},i_1}([e_{1_{p'}}] \wedge \mathbf{meet}_{p'}(e_1, x, y) \wedge \tau_{p'}(e_1) \supseteq i_1) \wedge$ $\rightsquigarrow_{Quantity2}$ $[e_1] \wedge \mathbf{meet}_{p^{cs}}(e_1, x, y) \wedge \tau_{p^{cs}}(e_1) \supseteq i_1$

b. Proposal: $[p] \wedge p \subseteq p^{cs} \wedge [i_2] \wedge [s_1] \wedge \mathbf{happy}_p(s_1, x) \wedge \tau_p(s_1) \supseteq i_2$

In short, since episodic readings can be obtained via pragmatic enrichment for zero-marked sentences in those narratives, no incompleteness arises.

Finally, for a zero-marked sentence with focus as in (48B), we've shown that it is normally uttered in a context in which it is already taken for granted that there is an event of Lisi meeting someone (e_1) completed within i_1 (49a). While the zero-marked imperfective sentence does not assert the occurrence of the event (49b), we can not only infer by default that the event of John meeting Mary has occurred within the topic time but also that this event is the same event as e_1. Via this kind of reasoning, we can see that the at-issue proposal put forth by B's utterance in fact concerns identifying the theme of e_1 as Mary (49c).

(48) A: zuotian Lisi jian-le yi-ge ren.
yesterday Lisi meet-PERF one-CL person
'Yesterday Lisi meet a person'

B: zuotian Lisi $\varnothing_{\text{IMPF}}$ jian MALI.
yesterday Lisi meet Mary
'It is $[\text{Mary}]_F$ Lisi met yesterday'

(49) Update the CS with the focused sentence *zuotian Lisi jian MALI*

a. Already satisfied by the CS: $[i_1] \wedge i_1 \subseteq \mathbf{yesterday} \wedge [x] \wedge x = \mathbf{Lisi} \wedge [y] \wedge \mathbf{person}_{p^{cs}}(y) \wedge [e_1] \wedge \mathbf{meet}_{p^{cs}}(e_1, x, y) \wedge \tau_{p^{cs}}(e_1) \subseteq i_1 \wedge$

b. Proposal: $[p] \wedge p \subseteq p^{cs} \wedge [i_2] \wedge [z] \wedge z = \mathbf{Mary} \wedge \mathbf{INERT}^{p'}_{p,i_2}([e_{2_{p'}}] \wedge \mathbf{meet}_{p'}(e_2, x, z) \wedge \tau_{p'}(e_2) \supseteq i_2) \wedge$

c. $\rightsquigarrow_{Quantity2}$ $[e_2] \wedge \mathbf{meet}_p(e_2, x, z) \wedge \tau_p(e_2) \supseteq i_2$ (e_2 accommodated)
$\rightsquigarrow e_1 = e_2$ (anaphoric resolution)
$\rightsquigarrow y = z = \mathbf{Mary}$

Since the occurrence of the event is also not directly relevant to the QUD (i.e. it is presupposed), there is no mandatory Quantity-1 implicature, and the pragmatic enrichment in (49c) is not blocked.

4.3 Interim Summary

In sum, the constrained, QUD-sensitive use of zero-marked sentences (i.e. imperfective forms) for episodic readings can be attributed to the interaction between two opposing sub-maxims under the Maxim of Quantity: Quantity-2 'Say no more than you need' and Quantity-1 'Say as much as you can (if relevant)'. The zero-marked form can be strengthened into episodic interpretations via the former sub-maxim, but such strengthening is not always successful because of the mandatory consideration of the better alternatives of expressing episodic readings when the occurrence of the event is at-issue in the discourse, due to the latter sub-maxim.

5 Conclusion

This paper investigated the incompleteness puzzle in Mandarin, namely zero-marked eventive sentences often sound degraded for episodic readings, with a novel pragmatic perspective. Inspired by the existing observations that incompleteness is not a context-free, across-the-board constraint [37,45,46,54], I established the novel generalization that incompleteness arises only when the occurrence of the event is (part of) the at-issue proposal put forth by an utterance. The generalization is motivated by the new data on clause-embedding eventive predicates in Mandarin, and can capture the existing observations that sentences within narratives or focused sentences generally do not suffer from incompleteness. Furthermore, I argued that Mandarin zero-marked sentences are in fact imperfective sentences since they can express typical imperfective readings, and it is not uncommon cross-linguistically for imperfective forms to express episodic readings in a constrained way. I proposed the first formal pragmatic account to capture the discourse-sensitivity of incompleteness, which involves an interaction between two sub-maxims under the Gricean Maxim of Quantity.

The current study raises a lot of open questions for future investigation. One concerns the cross-linguistic variation in the constraints of using imperfective forms for episodic readings (i.e. event-in-progress/event-completion readings). Previous studies have shown that the episodic use of the imperfective form can be blocked or constrained due to the availability of the more specific progressive or perfective markers in a language [9,17] and [21] shows in Russian how information structure regulates the aspectual competition between the perfective form and the imperfective form for expressing event-completion readings. It will be interesting to see how the aspectual competition between the zero-marked imperfective form and the progressive/perfective marker in Mandarin, which is regulated by what is at-issue in the discourse, can fit into this big picture. Another question is whether a similar pragmatic approach can be pursued for the other kinds of "incomplete" sentences in Mandarin such as (50). While this paper remains agnostic about whether different kinds of incompleteness should receive a uniform analysis or not, the fact that (50) can also improve with focus [18,38,39] suggests a pragmatic perspective can be promising.

(50) ??Lisi gao.
Lisi tall
Int: 'Lisi is tall'

(51) Lisi GAO, Mali SHOU.
Lisi tall Mary slim
'$[\text{Lisi}]_{CT}$ is $[\text{tall}]_F$, $[\text{Mary}]_{CT}$ is $[\text{slim}]_F$'

Acknowledgements. I would like to thank Itamar Francez, Anastasia Giannakidou, Dawei Jin, Chris Kennedy, Jackie Lai, Ming Xiang, Shumian Ye, Xuetong Yuan, and the anonymous reviewers of TLLM for comments and discussions. All remaining errors are mine.

References

1. Abrusán, M.: Predicting the presuppositions of soft triggers. Linguist. Philos. **34**, 491–535 (2011)
2. AnderBois, S., Brasoveanu, A., Henderson, R.: Crossing the appositive/at-issue meaning boundary. In: Semantics and Linguistic Theory, vol. 20, pp. 328–346 (2010)
3. AnderBois, S., Brasoveanu, A., Henderson, R.: At-issue proposals and appositive impositions in discourse. J. Semant. **32**(1), 93–138 (2015)
4. Arregui, A., Rivero, M.L., Salanova, A.: Cross-linguistic variation in imperfectivity. Nat. Lang. Linguist. Theory **32**(2), 307–362 (2014)
5. Atlas, J.D., Levinson, S.C.: It-clefts, informativeness and logical form: radical pragmatics (revised standard version). In: Radical Pragmatics, pp. 1–62. Academic Press (1981)
6. Bar-el, L., Davis, H., Matthewson, L.: On non-culminating accomplishments. In: Proceedings of NELS, vol. 35 (2005)
7. Chang, V.W.C.: The particle le in Chinese narrative discourse: an integrative description. Ph.D. thesis, University of Florida (1986)
8. Cremers, A., Coppock, L., Dotlačil, J., Roelofsen, F.: Ignorance implicatures of modified numerals. Linguist. Philos. **45**, 1–58 (2021)
9. Deo, A.: Unifying the imperfective and the progressive: partitions as quantificational domains. Linguist. Philos. **32**(5), 475–521 (2009)
10. Dowty, D.: Toward a semantic analysis of verb aspect and the English 'imperfective' progressive. Linguist. Philos. **1**, 45–77 (1977)
11. Dowty, D.R.: Word Meaning and Montague Grammar: The Semantics of Verbs and Times in Generative Semantics and in Montague's PTQ, vol. 7. Reidel, Dordrecht (1979)
12. Enç, M.: Anchoring conditions for tense. Linguist. Inq. 633–657 (1987)
13. Farkas, D.F., Bruce, K.B.: On reacting to assertions and polar questions. J. Semant. **27**(1), 81–118 (2010)
14. Feng, S.: Interactions between morphology syntax and prosody in Chinese. Beijing da xue chu ban she (1997)
15. Ferreira, M.: Event quantification and plurality. Ph.D. thesis, Massachusetts Institute of Technology (2005)
16. Ferreira, M.: The semantic ingredients of imperfectivity in progressives, habituals, and counterfactuals. Nat. Lang. Seman. **24**(4), 353–397 (2016). https://doi.org/10.1007/s11050-016-9127-2
17. Fuchs, M.: On the synchrony and diachrony of the Spanish imperfective domain: contextual modulation and semantic change. Ph.D. thesis, Yale University (2020)
18. Grano, T.: Mandarin hen and universal markedness in gradable adjectives. Nat. Lang. Linguist. Theory **30**(2), 513–565 (2012)
19. Grice, H.P.: Logic and conversation. In: Grice, P. (ed.) Studies in the Way of Words, pp. 41–58. Harvard University Press (1967)
20. Groenendijk, J., Stokhof, M.: Dynamic predicate logic. Linguist. Philos. 39–100 (1991)
21. Grønn, A.: The semantics and pragmatics of the Russian factual imperfective. Ph.D. thesis, University of Oslo, Oslo (2004)
22. Gu, Y.: Shitai, shizhi lilun yu hanyu shijian canzhao [studies of tense, aspect and Chinese time reference]. Yuyan kexue [Linguistic Sciences], pp. 22–38 (2007)

23. Guo, R.: Time reference and its syntactic effects in mandarin Chinese. Chin. Teach. World **29**(4), 435–449 (2015)
24. He, Y.: Time in mandarin: the fingerprints of tense and finiteness. Ph.D. thesis, Harvard University (2020)
25. Horn, L.: Towards a new taxonomy for pragmatic inference: Q-and R-based implicature. In: Shiffrin, D. (ed.) Meaning, Form and Use in Context, pp. 11–42. Georgetown University Press (1984)
26. Jayez, J.: Imperfectivity and progressivity: the French imparfait. In: Semantics and Linguistic Theory, vol. 9, pp. 145–162 (1999)
27. Katzir, R.: Structurally-defined alternatives. Linguist. Philos. **30**(6), 669–690 (2007)
28. Klein, W., Li, P., Hendriks, H.: Aspect and assertion in Mandarin Chinese. Nat. Lang. Linguist. Theory **18**(4), 723–770 (2000)
29. Koev, T.: Notions of at-issueness. Linguist. Lang. Compass **12**, e12306 (2018). https://doi.org/10.1111/lnc3.12306
30. Kong, L.: Yingxiang Hanyu juzi zizu de yuyanxingshi [Linguistic forms that affect sentence completeness in Chinese]. Zhongguo Yuwen **6**, 434–440 (1994)
31. van Kuppevelt, J.: Discourse structure, topicality and questioning. J. Linguist. **31**, 109–147 (1995)
32. van Kuppevelt, J.: Inferring from topics: scalar implicatures as topic-dependent inferences. Linguist. Philos. **19**, 393–443 (1996)
33. Landman, F.: The progressive. Nat. Lang. Seman. **1**(1), 1–32 (1992)
34. Lin, J.W.: Tenselessness. In: Binnick, R.I. (ed.) The Oxford Handbook of Tense and Aspect, pp. 669–695. Oxford University Press, Oxford (2012)
35. Liu, Z.: A syntax-semantics interface study of incompleteness effects in Chinese. Ph.D. thesis, The Chinese University of Hong Kong (2018)
36. Lin, J.W.: Aspectual selection and negation in Mandarin Chinese. Linguistics **41**(3), 425–459 (2003)
37. Lin, J.W.: Time in a language without tense: the case of Chinese. J. Semant. **23**(1), 1–53 (2006)
38. Liu, C.S.L.: The positive morpheme in Chinese and the adjectival structure. Lingua **120**(4), 1010–1056 (2010)
39. Liu, C.-S.L.: Projecting adjectives in Chinese. J. East Asian Linguis. **27**(1), 67–109 (2018). https://doi.org/10.1007/s10831-018-9166-4
40. Murray, S.E.: Varieties of update. Semant. Pragmat. **7**, 1–53 (2014)
41. Roberts, C.: Information structure in discourse: towards an integrated formal theory of pragmatics. Semant. Pragmat. **5**, 1–69 (1996/2012)
42. Sauerland, U.: Scalar implicatures in complex sentences. Linguist. Philos. **27**(3), 367–391 (2004)
43. Simons, M.: Observations on embedding verbs, evidentiality, and presupposition. Lingua **117**(6), 1034–1056 (2007)
44. Simons, M., Beaver, D., Roberts, C., Tonhauser, J.: The best question: explaining the projection behavior of factives. Discourse Process. **54**(3), 187–206 (2017)
45. Smith, C.S.: The Parameter of Aspect, vol. 43. Springer, Dordrecht (1997). https://doi.org/10.1007/978-94-011-5606-6
46. Smith, C.S., Erbaugh, M.S.: Temporal interpretation in Mandarin Chinese. Linguistics **43**(4), 713–756 (2005)
47. Sun, H.: Temporal construals of bare predicates in Mandarin Chinese. Ph.D. thesis, Leiden University dissertation (2014)
48. Sun, Y.: Incompleteness under discussion. Ph.D. thesis, The University of Chicago (2022)

49. Sybesma, R.: Xiandingxing he hanyu zhuju [Finiteness and Chinese main clauses]. Int. J. Chin. Linguist. **6**(2), 325–344 (2019)
50. Tang, S.W., Lee, T.H.T.: Focus as an anchoring condition. In: International Symposium on Topic and Focus in Chinese, The Hong Kong Polytechnic University (2000)
51. Tonhauser, J.: Diagnosing (not-) at-issue content. Proc. Semant. Under-Represented Lang. Am. (SULA) **6**, 239–254 (2012)
52. Tsai, W.T.D.: Tense anchoring in Chinese. Lingua **118**(5), 675–686 (2008)
53. Wang, C.: The syntax of le in Mandarin Chinese. Ph.D. thesis, Queen Mary University of London (2018)
54. Wu, J.S.: Tense as a discourse feature: rethinking temporal location in Mandarin Chinese. J. East Asian Linguis. **18**(2), 145–165 (2009)
55. Zhang, A.: On non-culminating accomplishments in Mandarin. Ph.D. thesis, The University of Chicago (2018)

Comparing the Derivation of Modal Domains and Strengthened Meanings

Tue Trinh(✉)

Leibniz-Zentrum Allgemeine Sprachwissenschaft, Berlin, Germany
trinh@leibniz-zas.de

Abstract. The derivation of strengthened meanings as proposed by Bar-Lev and Fox (2017, 2020) and the derivation of modal domains as proposed by Kratzer (1977, 1981, 1991) both involve an "inclusion" step of assigning *true* to as many propositions in a given set as possible. In the case of strengthened meanings, this set contains the scalar alternatives. In the case of modal domains, it contains the propositions in the ordering source. In this note, we explicate what is common and what is distinct between the two inclusion procedures. We then point out that the formal distinction makes no empirical difference for the cases of strengthened meaning so far considered in the literature. We conjecture that this fact holds generally for all cases of strengthened meaning.

Keywords: Modality · Exhaustification · Innocent inclusion · Cell identification · Alternatives

1 Two Steps of Exhaustification

1.1 Exclusion

The "grammatical approach to implicatures" takes the strengthened meaning of a sentence p, i.e. the conjunction of p and its implicatures, to result from applying an exhaustivity operator exh to p (cf. Fox 2007; Chierchia et al. 2012). Fox (2007) proposes that $exh(p)$ assigns *true* to p, the "prejacent", and assigns *false* to each of the "innocently excludable alternatives", henceforth "IE alternatives", of p.[1] We present Fox's 2007 proposal in (1), where A_p^{IE} is the set of IE alternatives of p and $\bigvee S$ is the proposition that at least one member of S is true, for any

This work is supported by the ERC Advanced Grant "Speech Acts in Grammar and Discourse" (SPAGAD), ERC-2007-ADG 787929.

[1] More precisely, $exh(p)$ assigns *true* to p and assigns *false* to each of the IE alternatives of p which are *relevant*. For the purpose of this discussion, we will make the simplifying assumption that the alternatives are all relevant. We do not believe this assumption affects our argument..

D. Deng et al. (Eds.): TLLM 2022, LNCS 13524, pp. 156–166, 2023.
https://doi.org/10.1007/978-3-031-25894-7_8

set S of propositions.[2] Innocent exclusion is defined in (2), where A_p is the set of alternatives of p.

(1) Fox's (2007) proposal
  a. The strengthened meaning of p is expressed by $exh(p)$
  b. $exh(p) \Leftrightarrow_{\text{def}} p \wedge \left(A_p^{IE} \neq \emptyset \rightarrow \neg \bigvee A_p^{IE}\right)$

(2) Fox's (2007) definition of A_p^{IE}
  (i) Take all maximal sets of propositions from A_p which can be assigned *false* consistently with p
  (ii) $q \in A_p^{IE}$ iff q is in all such sets

To illustrate, consider the Venn diagram below. Let p be the prejacent and q, r, s, t and p itself be its alternatives.[3] Logical relations are represented spatially in the familiar way. Thus, we have $p \Rightarrow (s \vee r \vee t)$, $(s \wedge t) \Rightarrow \bot$, for example.

(3)

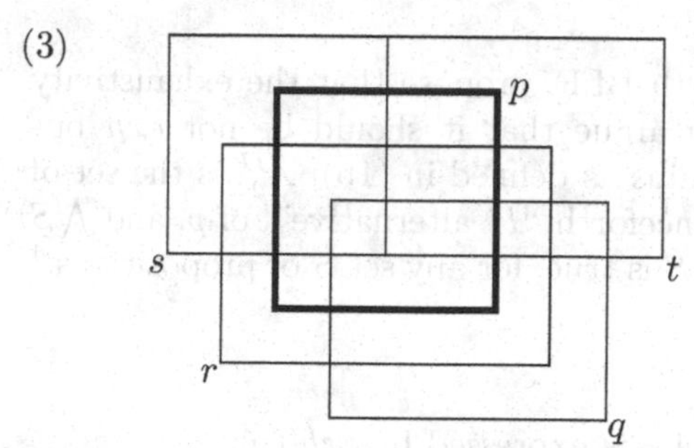

The maximal sets of propositions from A_p which can be assigned *false* consistently with p are listed in (4).

(4) a. $\{q, r, s\}$
  b. $\{q, r, t\}$
  c. $\{q, s, t\}$

Note that neither $\{r, s, t\}$ nor $\{q, r, s, t\}$ is listed, as $(\neg r \wedge \neg s \wedge \neg t) \Rightarrow \neg p$. Now, looking at (4), we see that only q is a member of all three sets. Thus, only q is an IE alternative of p, which means $exh(p) \Leftrightarrow p \wedge \neg q$. This proposition is indicated by the gray area in (5).

---

[2] The attentive reader will notice that the definition in (1b) contains a redundancy. Specifically, $A_p^{IE} \neq \emptyset \rightarrow \neg \bigvee A_p^{IE}$ is equivalent to $\neg \bigvee A_p^{IE}$, as $\bigvee \emptyset$ is the contradiction. The intuition which we want this redundant formulation to reflect is that the computation proceeds only under the condition that the relevant set of altermatives is not empty. That condition is logically idle for this case but not for all of the cases which we will discuss..

[3] We assume, as is standard, that every sentence is an alternative of itself (cf. Fox and Katzir 2011).

(5)

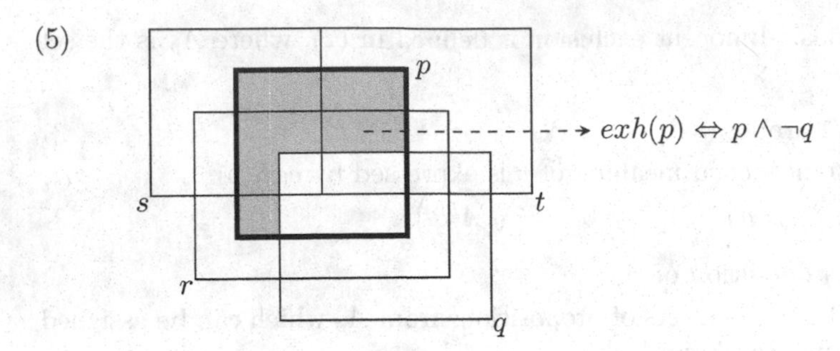

The process of exhaustification, as represented by *exh*, can therefore be described, informally, as that of trying to assign *false* to as many alternatives as possible, preserving consistency with the prejacent.

1.2 Inclusion

Bar-Lev and Fox (2017, 2020), henceforth BLF, propose that the exhaustivity operator be modified. Specifically, they argue that it should be not *exh* but *exh*′, as defined in (6b), where *exh* remains as defined in (1b), A_p^{II} is the set of "innocently *in*cludable alternatives", henceforth "II alternative", of p, and $\bigwedge S$ is the proposition that every member of S is true, for any set S of propositions.[4] Innocent inclusion is defined in (7).

(6) BLF's proposal

a. The strengthened meaning of p is expressed by $exh'(p)$

b. $exh'(p) \Leftrightarrow_{\text{def}} exh(p) \wedge \left(A_p^{II} \neq \emptyset \rightarrow \bigwedge A_p^{II}\right)$

(7) BLF's definition of A_p^{II}

(i) Take all maximal sets of propositions from A_p which can be assigned *true* consistently with $exh(p)$

(ii) $q \in A_p^{II}$ iff q is in all such sets

What $exh'(p)$ does, then, is assign *true* to $exh(p)$ and also assign *true* to each of the innocently includable alternatives of p.[5]. Consider, again, the Venn diagram in (3). Let us ask which among q, r, s, t and p itself is an II alternative of p. The maximal sets of propositions from A_p which can be assigned *true* consistently with $exh(p)$ are listed in (8).

(8) a. $\{p, r, s\}$

b. $\{p, r, t\}$

[4] Again, there is redundancy in (6b), as $\bigwedge \emptyset$ is the tautology. See note 2.

[5] Note that BLF claims that II alternatives are assigned *true* obligatorily (cf. Bar-Lev and Fox 2017, 111). Thus, the inferences associated with them cannot be cancelled by them being considered irrelevant, as is possible in the case of IE alternatives (see note 1)..

Note that neither $\{p, s, t\}$ nor $\{p, r, s, t\}$ is listed: since $s \wedge t$ is contradictory, no set containing s and t is consistent. Now, looking at the two sets in (8), we see that only p and r are members of both. This means only p and r are *II* alternatives of p, and that $exh'(p) \Leftrightarrow exh(p) \wedge p \wedge r \Leftrightarrow p \wedge \neg q \wedge r$. This proposition is indicated by the gray area in (9).

(9)

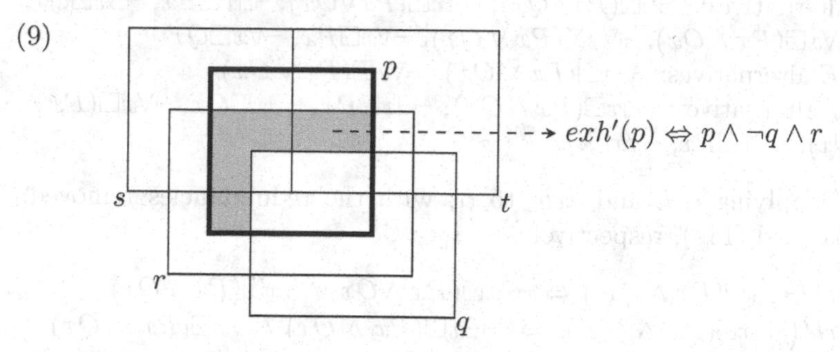

The process of exhaustification, as represented by exh', can therefore be described, informally, as that of (i) trying to assign *false* to as many alternatives as possible, preserving consistency with the prejacent, and then (ii) trying to assign *true* to as many alternatives as possible, preserving consistency with the output of (i). Thus, $exh'(p)$ is a strengthening of $exh(p)$. We can see this by comparing the gray area of (5) with the gray area of (9): the latter is a subpart of the former.

1.3 Empirical Motivation for *exh'*

BLF present a series of empirical arguments for identifying the strengthened meaning of p with $exh'(p)$ instead of $exh(p)$. Given the scope of this note, we will recite only one. The reader is invited to consult Bar-Lev and Fox (2017, 2020) to learn about the others.

The relevant data point is the sentence in (10), which has been argued to license the inferences in (10a) and (10b) (cf. Chemla 2009).

(10) No student is required to solve both problem A and problem B $\neg\exists x\Box(Px \wedge Qx)$

a. $\rightsquigarrow$ No student is required to solve problem A $\neg\exists x Px$

b. $\rightsquigarrow$ No student is required to solve problem B $\neg\exists x Qx$

The syntactic analysis of (10) at the relevant level, i.e. its Logical Form, is assumed to be something like (11).

(11) $[_\alpha$ no student λ_x $[_\beta$ is required to $[_\gamma$ t_x solve A and t_x solve B$]]]$

Given that **required** and **and** are both strong scalar items, exhaustifying β or γ will be semantically inconsequential.[6] The only scope site left for possibly non-vacuous exhaustification is the matrix node, which means the exhaustivity

[6] Because $exh(p \wedge q) = exh'(p \wedge q) = p \wedge q$, and $exh(\Box p) = exh'(\Box p) = \Box p$.

operator must be applied to α. BLF take the alternatives of α to be derived from α by replacing **no** ($\neg\exists$) with **not every** ($\neg\forall$), **and** with **or**, γ with its individual conjuncts, and α with itself.[7] We then have (12).

(12) a. Prejacent: $\neg\exists x\Box(Px \wedge Qx)$
b. Alternatives: $\neg\exists x\Box(Px\wedge Qx)$, $\neg\exists x\Box(Px\vee Qx)$, $\neg\exists x\Box Px$, $\neg\exists x\Box Qx$, $\neg\forall x\Box(Px \wedge Qx)$, $\neg\forall x\Box(Px \vee Qx)$, $\neg\forall x\Box Px$, $\neg\forall x\Box Qx$
c. IE alternatives: $\neg\exists x\Box(Px \vee Qx)$, $\neg\forall x\Box(Px \vee Qx)$
d. II alternatives: $\neg\exists x\Box(Px \wedge Qx)$, $\neg\exists x\Box Px$, $\neg\exists x\Box Qx$, $\neg\forall x\Box(Px \wedge Qx)$, $\neg\forall x\Box Px$, $\neg\forall x\Box Qx$

The results of applying exh and exh' to α, with the redundancies removed, amount to (13a) and (13b), respectively.

(13) a. $exh(\neg\exists x\Box(Px \wedge Qx)) \Leftrightarrow \neg\exists x\Box(Px \wedge Qx) \wedge \forall x\Box(Px \vee Qx)$
b. $exh'(\neg\exists x\Box(Px \wedge Qx)) \Leftrightarrow \neg\exists x\Box(Px \wedge Qx) \wedge \forall x\Box(Px \vee Qx) \wedge \neg\exists x\Box Px \wedge \neg\exists x\Box Qx$

We can see that the attested inferences can be derived with exh' but not with exh. More specifically, there is no way to derive these inferences with exh, but there is one way to derive them with exh'.[8]

2 A More Inclusive Inclusion

2.1 Conceptual Motivation for Inclusion

BLF mention a "possible underlying conception" which they say has "guided [their] thinking".

(14) Possible underlying conception (Bar-Lev and Fox 2020, 186)
Exhaustifying p with respect to a set of alternatives C should get us as close as possible to a cell in the partition induced by C

We quote from (Bar-Lev and Fox 2020, 186): "[...] [T]he goal of [the exhaustivity operator] is to come as close as possible to an assignment of a truth value to every alternative, i.e., to a cell in the partition that the set of alternatives induces

[7] The assumption that **no** ($\neg\exists$) alternates with **not every** ($\neg\forall$) is based on the analysis of **no** which decomposes it into **not** and **some** (cf. Zeijlstra 2004; Penka 2011), and on the view about alternative generation according to which negation is not replaced (cf. Romoli 2012). See Bar-Lev and Fox (2020, 198, note 32) on this point. Also, see Bar-Lev and Fox (2020, 198–200) for some independent reasons to assume that $\Box$ does not alternate with $\Diamond$ in this case.

[8] BLF point out that recursive application of exh does not help (cf. Bar-Lev and Fox 2020, 196). Note, also, that there is, in addition to the inferences in (10a) and (10b), another inference derived in (13b), namely $\forall x\Box(Px \vee Qx)$. This inference is really optional, since it should only arise if the alternative $\neg\forall x\Box(Px \vee Qx)$ is considered relevant, which it does not have to be. Again, we consider, for the purpose of this note, all alternatives to be relevant. (See note 1 and 5.).

[...] [The exhaustivity operator] is designed such that, when possible, it would yield a complete answer to the question formed by the set of alternatives. If this conception is correct, one would think that [it] shouldn't only exclude, i.e., assign *false* to as many alternatives as possible, but should also include, i.e., assign *true* to as many alternatives as possible once the exclusion is complete".

Looking at (5) and (9), we can see clearly how this idea plays out. The proposition expressed by $exh(p)$ consists of five cells in the partition induced by the alternatives, while the proposition expressed by $exh'(p)$ consists of three of these five cells. Thus, exhautification by exh' gets us closer to a single cell than exhaustification by exh.

2.2 Introducing exh''

Nothing in the definitions of innocent exclusion and innocent inclusion rules out the possibility of alternatives which are neither innocently excludable nor innocently includable. We will call such alternatives the "remaining alternatives", or "R alternatives" for short. Now let us entertain the hypothesis in (15), where A_p^R is the set of R alternatives of p and exh' is as defined in (6b).[9]

(15) Hypothesis

a. The strengthened meaning of p is expressed by $exh''(p)$

b. $exh''(p) \Leftrightarrow_{\text{def}} exh'(p) \wedge \left(A_p^R \neq \emptyset \rightarrow \bigvee A_p^R\right)$

(16) Definition of A_p^R

$q \in A_p^R$ iff $q \in A_p \wedge q \notin A_p^{IE} \wedge q \notin A_p^{II}$

The new exhaustivity operator we are considering, exh'', involves a more "inclusive" inclusion than exh'. Specifically, $exh''(p)$ not only includes the II alternatives by assigning *true* to each of them, but also "includes" the R alternatives by assigning *true* to their disjunction. Thus, $exh''(p)$ is a strengthening of $exh'(p)$, which means exh'' actually comes closer to BLF's "underlying conception" of exhaustification than exh'. We can see this by looking at (17), where the gray area represents $exh''(p)$.

[9] Note that the condition that A_p^R not be empty is significant here. We want to capture the intuition that if there is no R alternative, the system would just output $exh'(p)$. Specifically, we do want it to not output the contradiction in case A_p^R is empty, which is what would happen if $exh''(p)$ were defined as $exh'(p) \wedge \bigvee A_p^R$.

(17)

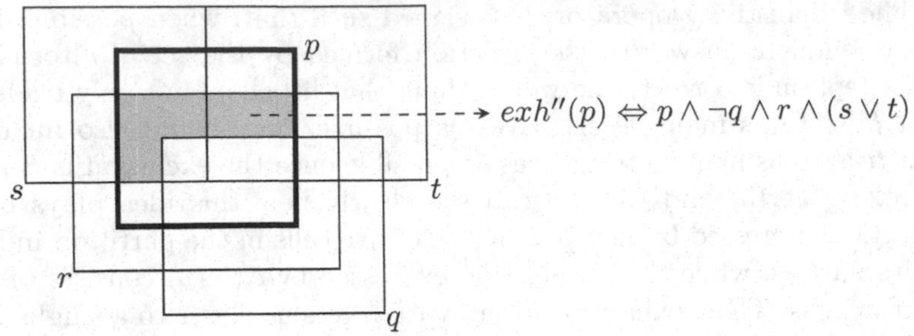

p

$exh''(p) \Leftrightarrow p \wedge \neg q \wedge r \wedge (s \vee t)$

s t

r

q

Comparing (17) to (9), we see that the proposition expressed by $exh'(p)$ consists of three cells and the proposition expressed by $exh''(p)$ consists of two of those three cells. Thus, $exh''(p)$ is closer to a complete answer of the question formed by the set of alternatives than $exh'(p)$.

2.3 A Resemblance

Kratzer (1977, 1981, 1991) propose that modality is "double relative". Specifically, the quantification domain D of a modal operator is specified in terms of two sets of propositions, a "modal base" M and an "ordering source" O,[10] in the following way.

(18) Derivation of D from M and O

a. Take all maximal sets of propositions from O which can be assigned *true* consistently with $\bigwedge M$

b. D is the result of conjoining $\bigwedge M$ with

(i) propositions that are in all such sets

(ii) the disjunction of the remaining propositions in O

A necessity statement $\Box a$ is then true iff a is entailed by D, and a possibility statement $\Diamond a$ is true iff a is consistent with D.

As we can see, the two steps (18b-i) and (18b-ii) resemble the inclusion of II and R alternatives, respectively. Thus, if we identify $\bigwedge M$ with $exh(p)$ and O with $A_p - A_p^{IE}$, then we can identify D with $exh''(p)$. Let us, again, use our Venn diagram to illustrate. Suppose $M = \{p, \neg q\}$ and $O = \{r, s, t\}$. Then D is the gray area, which corresponds to $exh''(p)$. Importantly, D is not the dotted area, which corresponds to $exh'(p)$.

[10] Technically, the two sets of propositions are values of the modal base and the ordering source at the evaluation world, as these are functions from worlds to sets of propositions. The reader is invited to consult Kratzer (1977, 1981, 1991) for a more precise and sophisticated presentation of her theory. Relevant secondary literature includes von Fintel and Heim (2011); Frank (1996), among others.

(19)

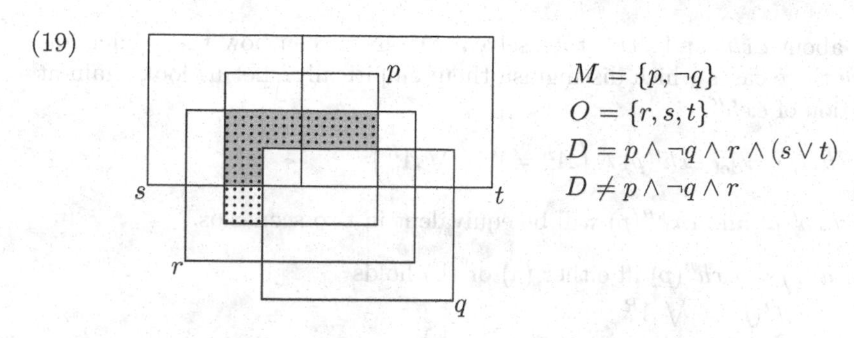

We will illustrate with an example. Let us give the following meanings to r, s, and t.[11]

(20) a. r = John volunteered as poll watcher
b. s = John voted Republican
c. t = John voted Democrat

And let it be common ground that $p \wedge \neg q$.[12] This will be the modal base. Suppose that John's father says he voted Republican (s), John's mother says he voted Democrat (t), and both of John's parents say he volunteered as poll watcher (r). This will be the ordering source. Now consider the following sentences.

(21) a. In view of what his parents say, it is possible that John volunteered as poll watcher and voted Republican $\diamond(r \wedge s)$
b. In view of what his parents say, it is possible that John volunteered as poll watcher and did not vote $\diamond(r \wedge \neg s \wedge \neg t)$

If $D = p \wedge \neg q \wedge r$, the dotted area, we expect both (21a) and (21b) to be true, as both $r \wedge s$ and $r \wedge \neg s \wedge \neg t$ are consistent with $p \wedge \neg q \wedge r$. If $D = p \wedge \neg q \wedge r \wedge (s \vee t)$, the gray area, we expect (21a) to be true and (21b) to be false, as $r \wedge s$ is consistent with $p \wedge \neg q \wedge r \wedge (s \vee t)$ but $r \wedge \neg s \wedge \neg t$ is not. Our intuition is that (21a) is true and (21b) is false. This fact constitutes evidence that D is $p \wedge \neg q \wedge r \wedge (s \vee t)$, the gray area, and not $p \wedge \neg q \wedge r$, the dotted area.

3 A Conjecture

The Kratzerian inclusion of the ordering source involves including the disjunction of propositions which are not "innocently includable". Thus, it resembles the inclusion step of *exh″*, not that of *exh′*. Our judgement about (21a) and (21b) confirms that Kratzer is correct.

[11] We will assume that one can vote for only one party, and that the only choices are Republican and Democrat.

[12] In other words, let what we know be consistent with r, s, t, $r \wedge s$, $r \wedge t$, and let it assymetrically entail $r \vee s \vee t$. For concreteness, we can take this body of information to be the proposition that John lived in D.C and either voted or volunteered as poll watcher.

What about exh' and exh'' themselves? We have seen how they differ formally. Are there cases which distinguish them empirically? Let us look again at the definition of exh''.

(22) $exh''(p) \Leftrightarrow_{\text{def}} exh'(p) \wedge \left(A_p^R \neq \emptyset \rightarrow \bigvee A_p^R\right)$

Logically, $exh'(p)$ and $exh''(p)$ will be equivalent in two scenarios.

(23) $exh'(p) \Leftrightarrow exh''(p)$ iff either (a) or (b) holds

a. $exh'(p) \Rightarrow \bigvee A_p^R$

b. $A_p^R = \emptyset$

Consider (23a) first. This scenario is instantiated by plain disjunctions such as (24).

(24) John talked to Mary or Sue ($p \vee q$)

a. Alternatives: $p \vee q, p, q, p \wedge q$

b. *IE* alternatives: $p \wedge q$

c. *II* alternatives: $p \vee q$

d. *R* alternatives: p, q

e. $exh'(p \vee q) \Leftrightarrow (p \vee q) \wedge \neg(p \wedge q)$

Now consider (23b). This scenario is exemplified by (10), discussed in Sect. 1.3. As the reader can see from (12), the *IE* and *II* alternatives of (10) exhaust the set of its alternatives. Thus, there are no *R* alternatives left.

Another case where every alternative is either *IE* or *II* is one involving the scalar items **all, many, some**. Consider the three sentences in (25).

(25) a. John did all of the homeworks

b. John did many of the homeworks

c. John did some of the homeworks

Each of these sentences has all three as alternatives. This means that for (25a), every alternative is *II*. For (25b), (25a) is *IE* while (25b) and (25c) are *II*. And for (25c), (25a) and (25b) are *IE* while (25c) is *II*.

When will $exh'(p)$ and $exh''(p)$ not be equivalent? Obviously when both (23a) and (23b) are false. Has a case been discussed in the literature which exemplifies this possibility? The answer to this question, we believe, is negative. As far as we know, all cases considered in the literature on exhaustification so far, including those discussed in Bar-Lev and Fox (2017, 2020), are either an instance of (23a) or an instance of (23b). We conjecture that this holds generally for all cases of exhaustification.

(26) Strengthened Meaning Conjecture (SMC)
There is no sentence p in natural language such that the strengthened meaning of p is $exh'(p)$ but not $exh''(p)$

From our discussion in Sects. 2.1 and 2.3, it is clear that the exhaustivity operator could in principle be exh'', not exh', and that the distinction between exh'' and exh' could in principle make an empirical difference. We could imagine the facts about semantic strengthening to be such that they adjudicate between the two different inclusion procedures involved in exhaustification, just as facts about modality do with respect to the ordering source. So what is missing? We believe that SMC will follow given a complete theory of alternatives. In other words, we believe that such a theory would rule out the scenario in (3) as a grammatical impossibility. We therefore formulate the following challenge for future research.

(27) Challenge
Construct the theory of alternatives so that SMC follows

4 Conclusion

Bar-Lev and Fox (2017, 2020) propose to add inclusion to exhaustification. In addition to providing empirical arguments for their proposal, they also note that the addition makes conceptual sense given the natural understanding of semantic strengthening as an attempt by the grammar to get as close as possible to a complete answer to the question formed by the set of alternatives. We discuss a variant of inclusion which would better represent this attempt than the variant proposed by Bar-Lev and Fox. We show that the new variant resembles the inclusion of ordering sources in Kratzer's (1977, 1981, 1991) theory of modality. We point out that the two variants end up being empirically equivalent for the cases of strengthened meaning so far considered in the literature. We conjecture that this equivalence is a general fact about exhaustification, and pose the challenge of deriving it for future research.

References

Bar-Lev, M., Fox, D.: Universal free choice and innocent inclusion. In: Proceedings of SALT, vol. 27, pp. 95–115 (2017)

Bar-Lev, M., Fox, D.: Free choice, simplification, and innocent inclusion. Nat. Lang. Seman. **28**, 175–223 (2020)

Chemla, E.: Universal implicatures and free choice effects: experimental data. Semant. Pragmat. **2** (2009)

Chierchia, G., Fox, D., Spector, B.: The grammatical view of scalar implicatures and the relationship between semantics and pragmatics. In: Portner, P., Maienborn, C., von Heusinger, K. (eds.) Semantics: An International Handbook of Natural Language Meaning, pp. 2297–2332. De Gruyter (2012)

von Fintel, K., Heim, I.: Intensional Semantics. MIT Lecture Notes, 2011 Edition (2011). https://mit.edu/fintel/fintel-heim-intensional.pdf

Fox, D.: Free choice disjunction and the theory of scalar implicatures. In: Sauerland, U., Stateva, P. (eds.) Presupposition and Implicature in Compositional Semantics, pp. 71–120. Palgrave-Macmillan (2007)

Fox, D., Katzir, R.: On the characterization of alternatives. Nat. Lang. Seman. **19**, 87–107 (2011)

Frank, A.: Context dependence in modal constructions. Doctoral Dissertation, Universität Stuttgart (1996)

Kratzer, A.: What "must" and "can" must and can mean. Linguist. Philos. **1**, 337–355 (1977)

Kratzer, A.: The notional category of modality. In: Eikmeyer, H., Rieser, H. (eds.) Words, Worlds, and Contexts: New Approaches in Word Semantics, pp. 38–74. De Gruyter (1981)

Kratzer, A.: Modality. In: von Stechow, A., Wunderlich, D. (eds.) Semantics: An International Handbook of Contemporary Research, pp. 639–650. De Gruyter (1991)

Penka, D.: Negative Indefinites. Oxford University Press, Oxford (2011)

Romoli, J.: Soft but Strong. Neg-raising, soft triggers, and exhaustification. Doctoral Dissertation, Harvard University (2012)

Zeijlstra, H.: Sentential Negation and Negative Concord. LOT, Utrecht (2004)

Post-suppositions and Uninterpretable Questions

Linmin Zhang[1,2](✉)

[1] NYU Shanghai, Shanghai, China
[2] NYU-ECNU Institute of Brain & Cognitive Science, Shanghai, China
zhanglinmin@gmail.com, linmin.zhang@nyu.edu
https://sites.google.com/site/zhanglinmin/

Abstract. For a sentence like *exactly three boys are between 5 ft 10 in. and 6 ft tall*, why cannot we abstract the height information out and raise a corresponding degree question like #*How tall are exactly three boys*? Inspired by the ideas that (i) there is a connection between *wh*-questions (e.g., *who did Mary kiss*) and definite descriptions (e.g., *the people that Mary kissed*) and (ii) definite descriptions and modified numerals (e.g., *exactly three boys*) bring post-suppositions (i.e., delayed evaluations that lead to relative definiteness, Brasoveanu 2013, Bumford 2017), I propose that when different elements that bring post-suppositions are present, a potential conflict arises in computing relative definiteness, leading to uninterpretability.

Keywords: Dynamic semantics · Post-suppositions · *Wh*-questions · Degree questions · Modified numerals · Cumulative reading · Definiteness · Weak island effects · Intervention effects

1 Introduction

This paper aims to explain the unacceptability of sentences like (2b): a **constituent question** containing a **modified numeral**.

(1) a. Brienne is <u>between 5′10″ and 6′</u> tall.
b. How tall is Brienne?

(2) a. Exactly three boys are <u>between 5′10″ and 6′</u> tall.
b. #How tall are exactly three boys?

For a sentence like (1a), we can naturally abstract the height information (the underlined part) out and raise a corresponding degree question on Brienne's

This project was financially supported by the Program for Eastern Young Scholars at Shanghai Institutions of Higher Learning (to L.Z.). The current paper supersedes an earlier manuscript included in the Pre-Conference *Proceedings of LENLS16* (Zhang 2019). I thank Anna Szabolcsi and the anonymous reviewers and audience of both LENLS16 and TLLM2022 for comments and feedback. Errors are mine.

D. Deng et al. (Eds.): TLLM 2022, LNCS 13524, pp. 167–187, 2023.
https://doi.org/10.1007/978-3-031-25894-7_9

height (see (1b)). Intriguingly, in contrast to (1), for a sentence like (2a), which contains a modified numeral (here *exactly three boys*), abstracting the height information out to form a corresponding degree question does not work, yielding an intuitively unacceptable sentence (see (2b)).

The uninterpretability of constituent questions like (2b) does not seem like an entirely new observation. Similar unacceptable question phenomena have been reported in the literature on **intervention effects** or **weak island effects** (see, e.g., Szabolcsi and Zwarts 1993, Szabolcsi 2006, Rullmann 1995, Honcoop 1996, Beck 1996, 2006, Fox and Hackl 2007, Abrusán 2014).

As illustrated by the contrast between (3a) and (3b) (i.e., *wh*-in-situ vs. *wh*-movement), **intervention effects** arise when an intervener (here the negation expression *koi nahiiN*) precedes a *wh*-word (here *kyaa*) in a *wh*-question. This kind of intervention effects are often attested in *wh*-in-situ languages (e.g., Hindi, Korean). Cross-linguistically, typical interveners include, but are not limited to, focus particles and downward entailing (DE) quantifiers (e.g., *no*, *few*, *at most*).

(3) **Intervention effects**: examples from Beck (2006)

a. ?? koi **nahiiN** kyaa paRhaa
anyone not what read-Perf.M
Intended: 'What did no one read?' (Hindi: (12a) in Beck 2006)

b. kyaa koi **nahiiN** paRhaa
what anyone not read-Perf.M
'What did no one read?' (Hindi: (12b) in Beck 2006)

Islands refer to domains which prevent the displacement of items contained within them, and **weak islands** are those that are only closed for some kinds of items, but not all kinds of items (see Szabolcsi 2006, Abrusán 2014). As illustrated by (4), negation words or DE quantifiers create **weak islands effects** in the formation of a degree question (see (4a) and (4c)), *how-many* question (see (4b)), or manner question (see (4d)). In contrast, negation words or DE quantifiers do not create islands for the displacement of items like *which book* (see (5)). Elements that create weak island effects are also not limited to negation operators or DE quantifiers.

(4) **Weak island effects**: examples from Abrusán (2014)

a. #How tall is**n't** John? (§3.4, (32a))
b. ??How many children does **none of these women** have? (§5.3, (19))
c. #How far did **few girls** jump? (§5.3, (24c))
d. #How did **at most 3 girls** behave? (§5.3, (24e)

(5) a. Which book have**n't** you read? (Abrusán 2014: §1.1, (3))
b. Which book did { **no one** / **few girls** / **at most 3 girls** } read?

Within the existing literature, there are already a variety of proposals on intervention effects or weak island effects, sometimes with different empirical coverages. The pattern 'modified numeral + degree question' (see (2b)) seems relevant, but it has not been much studied as a core piece of data. In this paper, I propose to start with the special property of modified numerals that they bring post-suppositions (Brasoveanu 2013) and explore how far this new perspective can advance our understanding of sentences like (2b) as well as empirical data related to intervention effects or weak island effects.

In a nutshell, I adopt and develop existing ideas in the literature on *wh*-questions: there is a connection between the interpretation of *wh*-questions (e.g., *who did Mary kiss*) and definite descriptions (e.g., *the people that Mary kissed*). Then given that definite descriptions and modified numerals are both elements that bring **post-suppositions** (see Brasoveanu 2013, Bumford 2017), i.e., delayed evaluations that result in a deterministic update with relative definiteness, the presence of both these kinds of items in the same sentence potentially yields a conflict with regard to relative definiteness, leading to uninterpretability. I will also address how this potential uninterpretability can be circumvented.

In the following, Sect. 2 first presents how modified numerals and definite descriptions contribute post-suppositions (Brasoveanu 2013, Bumford 2017). Section 3 argues for a parallel analysis for interpretable *wh*-questions and modified numerals / definite descriptions. Based on this, Sect. 4 accounts for the uninterpretability of the core data under discussion (see (2b)). Section 5 compares the current proposal with existing approaches developed within the literature on intervention effects and weak island effects and shows advantages of the current proposal. Section 6 concludes.

2 Post-suppositions

2.1 Brasoveanu (2013): Modified Numerals as Post-suppositions

Modified numerals bring **post-suppositions**: their numerical information is attached to a **non-local, sentence-level maximization** (Brasoveanu 2013).

The maximization effect of modified numerals has been widely reported in the literature (see, e.g., Szabolcsi 1997, Krifka 1999, de Swart 1999, Umbach 2006, Zhang 2018). As illustrated by the contrast in (6), compared to bare numerals like *two dogs*, modified numerals like *at least two dogs* exhibit maximality, as evidenced by the infelicitous continuation *perhaps she fed more.* In other words, while the semantic contribution of *two* in (6a) is **existential**, *at least two* in (6b) conveys the quantity information of the **totality** of dogs fed by Mary.

(6) a. Mary fed <u>two dogs</u>. They are cute. Perhaps she fed <u>more</u>.
  b. Mary fed <u>at least two dogs</u>. They are cute. #Perhaps she fed <u>more</u>.

The non-localness of this maximization is best reflected in the **cumulative reading** of sentences like (7). (7) has a distributive reading (7a) and a cumulative reading (7b), and we focus on the cumulative reading (7b) here. (For notation simplicity, cumulative closure is assumed for lexical relations when needed.)

(7) Exactly 3 boys saw exactly 5 movies.

a. **Distributive reading:**

$$\underbrace{\sigma x[\text{BOY}(x) \wedge \delta x[\underbrace{\sigma y[\text{MOVIE}(y) \wedge \text{SEE}(x,y)]}_{\text{the mereologically maximal } y} \wedge |y| = 5]]}_{\text{the mereologically maximal } x} \wedge |x| = 3$$

(σ: maximality operator; δ: distributivity operator.)
(There are in total three boys, and for each atomic boy, there are in total 5 movies such that he saw them.)

b. **Cumulative reading:**

$$\underbrace{\sigma x \sigma y[\text{BOY}(x) \wedge \text{MOVIE}(y) \wedge \text{SEE}(x,y)]}_{\text{the mereologically maximal } x \text{ and } y} \wedge |y| = 5 \wedge |x| = 3$$

(The cardinality of all the boys who saw any movies is 3, and the cardinality of all movies seen by any boys is 5.)
True under the context of Fig. 1, **false** under the context of Fig. 2.

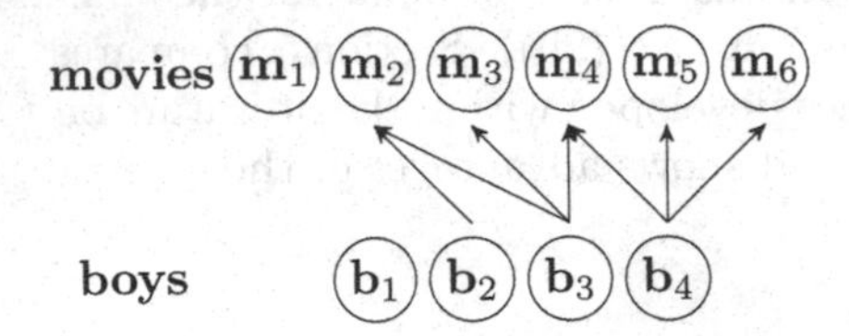

Fig. 1. The **cumulative** reading of *exactly 3 boys saw exactly 5 movies* is **true** under this scenario.

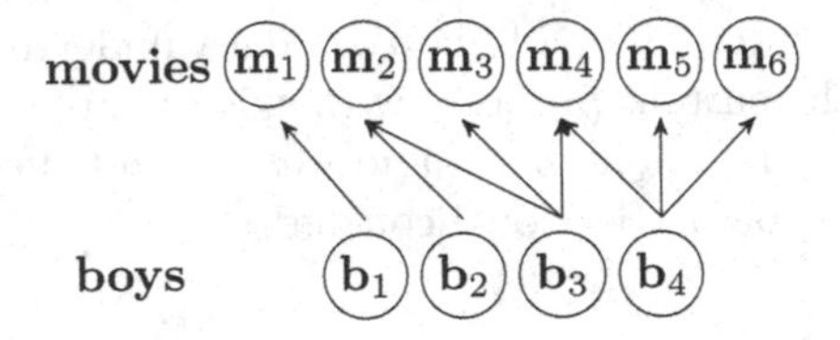

Fig. 2. The **cumulative** reading of *exactly 3 boys saw exactly 5 movies* is **false** under this scenario.

According to the intuition of native speakers, sentence (7) is true under the scenario described by Fig. 1, but false under the scenario described by Fig. 2.

It is worth noting that if we adopt the analysis shown in (8), then sentence (7) should be judged true under the scenario of Fig. 2: there are two such boy-sum witnesses, namely $b_2 \oplus b_3 \oplus b_4$ and $b_1 \oplus b_2 \oplus b_4$, and for each of these two boy-sums, (i) their cardinality is 3, and (ii) the maximal sum of movies seen between them has the cardinality of 5 ($m_2 \oplus m_3 \oplus m_4 \oplus m_5 \oplus m_6$ and $m_1 \oplus m_2 \oplus m_4 \oplus m_5 \oplus m_6$, respectively). There are no larger boy-sums such that they saw in total 5 movies between them. Thus the contrast of intuition (i.e., (7) is true under Fig. 1, but false under Fig. 2) means that (i) the genuine cumulative reading shown in (7b) is distinct from the unattested pseudo-cumulative reading shown in (8), and (ii) there is no scope-taking between the two modified numerals in (7), *exactly 3 boys* and *exactly 5 movies* (see Brasoveanu 2013, Charlow 2017).

(8) **Unattested pseudo-cumulative reading** of (7): Not attested!

$$\underbrace{\sigma x[\textsc{boy}(x) \wedge \underbrace{\sigma y[\textsc{movie}(y) \wedge \textsc{see}(x,y)]}_{\text{the mereologically maximal } y} \wedge |y| = 5]}_{\text{the mereologically maximal } x} \wedge |x| = 3$$

(The maximal plural individual x satisfying the restrictions (i.e., atomic members of x are boys, each atomic boy saw some movies, and the boys in x saw a total of 5 movies between them) has the cardinality of 3.) **True** under the context of Fig. 2 (see $b_2 \oplus b_3 \oplus b_4$ and $b_1 \oplus b_2 \oplus b_4$)!

As already pointed out by Krifka (1999), the semantic contribution of both modified numerals in (7), *exactly 3 boys* and *exactly 5 movies*, should take place simultaneously, at the sentential level, beyond their hosting DPs themselves:

> The problem cases discussed here clearly require a representation in which NPs are not scoped with respect to each other. Rather, they ask for an interpretation strategy in which all the NPs in a sentence are somehow interpreted on a par. (Krifka 1999)

Given Fig. 1, in interpreting (7), we count the cardinalities of all boys who saw any movies and all movies seen by any boys, instead of the total cardinalities of all boys and movies in the domain (here in Fig. 1, it's 4 boys and 6 movies). Therefore, the application of maximality operators is subject to more restrictions (here in our context, not just boys, but boy who saw movies; not just movies, but movies seen by boys), leading to a **relativized maximization** effect.

A compositional analysis à la Bumford (2017) is sketched in (9). Within dynamic semantics, meaning derivation is considered a series of updates from an information state to another. The semantic contribution of modified numerals is split. They first introduce discourse referents (drefs), x and y. Restrictions like $\textsc{movie}(y)$, $\textsc{boy}(x)$, and $\textsc{saw}(x, y)$ are added onto these drefs. Eventually, it is after all these restrictions are applied that maximality and cardinality tests, $\mathbf{M}_u/\mathbf{M}_\nu/\mathbf{3}_u/\mathbf{5}_\nu$, as delayed evaluations, i.e., post-suppositional tests, come into force. $\mathbf{M}_u$ and $\mathbf{M}_\nu$ check whether u and ν are assigned the mereologically maximal plural individuals x and y that satisfy all the restrictions, and $\mathbf{3}_u$ and $\mathbf{5}_\nu$ check whether the cardinalities of maximal x and y are 3 and 5 respectively.

(9) $\lambda g.\left\{\left\langle T, g^{\nu \mapsto y}_{u \mapsto x}\right\rangle \,\middle|\, \begin{array}{l} y = \sigma y.[\textsc{movie}(y) \wedge \exists x.[\textsc{boy}(x) \wedge \textsc{saw}(x,y)]] \\ x = \sigma x.[\textsc{boy}(x) \wedge \exists y.[\textsc{movie}(y) \wedge \textsc{saw}(x,y)]] \end{array}\right\}$, if $|x| = 3$ and $|y| = 5$

$3_u \bullet 5_\nu$ (the definite part of **exactly N_u** and **exactly N_ν**)

$\lambda g.\left\{\left\langle T, g^{\nu \mapsto y}_{u \mapsto x}\right\rangle \,\middle|\, \begin{array}{l} y = \sigma y.[\textsc{movie}(y) \wedge \exists x.[\textsc{boy}(x) \wedge \textsc{saw}(x,y)]] \\ x = \sigma x.[\textsc{boy}(x) \wedge \exists y.[\textsc{movie}(y) \wedge \textsc{saw}(x,y)]] \end{array}\right\}$

$M_u \bullet M_\nu$ (the definite part of **exactly N_u** and **exactly N_ν**)

$\lambda g.\left\{\left\langle T, g^{\nu \mapsto y}_{u \mapsto x}\right\rangle \,\middle|\, \begin{array}{l} \textsc{movie}(y), \\ \textsc{boy}(x), \\ \textsc{saw}(x,y) \end{array}\right\}$

$\lambda g.\,\{\langle x, g^{u \to x}\rangle \mid \textsc{boy}(x)\}$

some^u (the indefinite part of **exactly N_u**) boys

saw $\lambda g.\,\{\langle y, g^{\nu \mapsto y}\rangle \mid \textsc{movie}(y)\}$

some$^\nu$ (the indefinite part of **exactly N_ν**) movies

(Here $\mathbf{M}_\nu \overset{\text{def}}{=} \lambda m.\lambda g.\,\{\langle \alpha, h\rangle \in m(g) \mid \neg\exists\langle \beta, h'\rangle \in m(g).\, h(\nu) \sqsubset h'(\nu)\}$.[1]
$\bullet$ is used to simplify the notation in bundling two tests together.)

2.2 Bumford (2017): Definite Descriptions as Post-suppositions

Not only modified numerals bring post-suppositions, Bumford (2017)'s analysis for Haddock (1987)'s example (see Fig. 3) shows that **definite descriptions** like *the rabbit in the hat* also involve post-suppositions, i.e., delayed tests that lead to **relativized definiteness** effects.

Under the scenario shown in Fig. 3, there are multiple rabbits (R1, R2, R3) and multiple hats (H1, H2). Thus, the uniqueness requirement of *the rabbit* or *the hat* cannot be met in an absolute sense. However, *the rabbit in the hat* is still perfectly felicitous in this context.

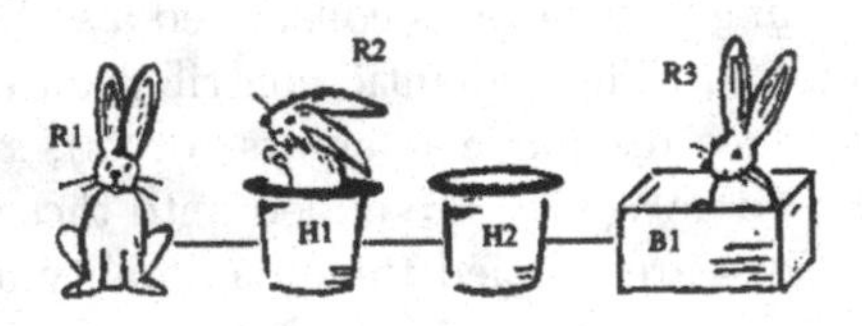

Fig. 3. The rabbit in the hat

Bumford (2017) argues that Haddock (1987)'s definite description is exactly parallel to the case of *exactly 3 boys saw exactly 5 movies*, where maximality tests are applied on drefs satisfying all these restrictions including $\textsc{movie}(y)$, $\textsc{boy}(x)$, and $\textsc{saw}(x,y)$, resulting in relativized maximization.

[1] The type of $\mathbf{M}_\nu$ is $(g \to \{\langle\alpha, g\rangle\}) \to (g \to \{\langle\alpha, g\rangle\})$, with g meaning the type for assignment functions, and α standing for the type of the denotation corresponding to the constituent. The usual notation for types $\langle\alpha, \beta\rangle$ is written as $\alpha \to \beta$ here.

As shown in (10), under the given scenario in Fig. 3, for *the rabbit in the hat*, uniqueness tests $\mathbf{1}_\nu/\mathbf{1}_u$ are also applied in a delayed, non-local manner, after the introduction of all the drefs (i.e., x and y) and restrictions (i.e., $\text{HAT}(y)$, $\text{RABBIT}(x)$, and $\text{IN}(x, y)$). More specifically, the test $\mathbf{1}_\nu$ first checks whether there is a unique hat in the context such that only this hat contains any rabbits. Then the test $\mathbf{1}_u$ checks whether the rabbit contained in the above-mentioned unique rabbit-containing hat is unique.[2]

(10) The^u rabbit in the$^\nu$ hat $\rightsquigarrow$ rabbit R2 in Figure 3

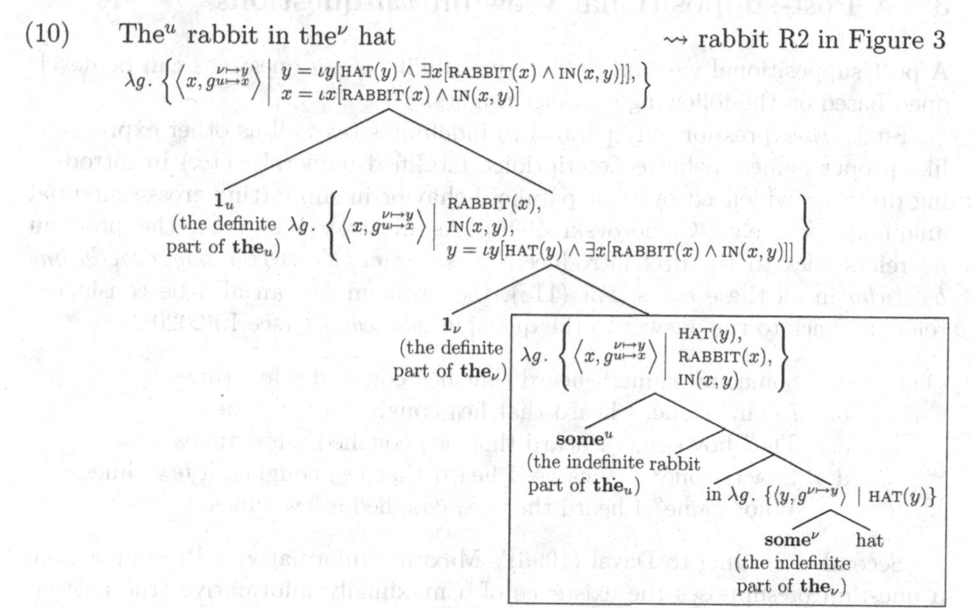

The upshot is that the semantic contribution of modified numerals and definite descriptions can be considered split, (i) introducing drefs at an earlier stage,

[2] An anonymous reviewer questions about how we can decide the order of different maximality tests and whether the order of maximality tests in the trees of (9) and (10) indicate scopal relations.

For the order of applying maximality tests in (9) and (10), I basically follow the original analysis of Brasoveanu (2013) and Bumford (2017).

For (9), according to Brasoveanu (2013), the maximality tests on the two plural individual drefs need to be applied **simultaneously**, as delayed tests (i.e., after the introduction of all relevant drefs and the application of relevant restrictions). The simultaneous application of the two maximality tests is intended for the non-scopal cumulative reading (see also Krifka 1999's discussion, and see Zhang 2022's discussion on attributing this simultaneity to contextually relevant QUD).

For (10), I also follow Bumford (2017) in applying the two uniqueness tests one after the other. In my view, these two uniqueness tests can also be applied simultaneously in picking out the unique rabbit-hat pair.

The order of applying maximality tests does have an effect similar to QR-styled scope-taking, but via a distinct mechanism. A more detailed discussion is beyond the scope of the current paper (see Charlow 2017).

and (ii) then at a later stage, imposing delayed, post-suppositional tests and leading to a relativized maximization/definiteness effect.

Moreover, modified numerals (e.g., *exactly 3 boys*) are distinct from definite descriptions (e.g., *the 3 boys*) in that the former can only be relatively definite, while the latter can be either absolutely or relatively definite.[3]

3 A Post-suppositional View on *wh*-questions

A post-suppositional view on the interpretability of *wh*-questions can be developed based on the following existing insights.

First, *wh*-expressions are parallel to indefinites (as well as other expressions like proper names, definite descriptions, modified numerals, etc.) in introducing drefs, as evidenced by their parallel behavior in supporting cross-sentential anaphora (see, e.g., Comorovski 1996). As illustrated in (11), the pronoun *he* refers back to the dref introduced by *someone/Kevin/the boy/exactly one boy/who* in all these cases. For (11e), the pronoun *he* can also be considered referring back to the answer to the question *who came?* (see Li 2020).

(11) a. Someone^0 came. I heard that he$_0$ coughed a few times.
b. Kevin^0 came. I heard that he$_0$ coughed a few times.
c. The^0 boy came. I heard that he$_0$ coughed a few times.
d. Exactly one^0 boy came. I heard that he$_0$ coughed a few times.
e. Who^0 came? I heard that he$_0$ coughed a few times.

Second, according to Dayal (1996)'s Maximal Informativity Presupposition, a question presupposes the existence of a maximally informative true answer. This idea can be combined with the Hamblin-Karttunen semantics of questions to reason about the (non-)deterministic updates of propositions.

According to Hamblin (1973), a *wh*-question denotes a set of propositions, which are **possible propositional answers** to the question. Then according to Karttunen (1977), a *wh*-question denotes the set of its **true propositional answers**. As illustrated in (12), we can use an answerhood operator to bridge the set of possible answers and the maximally informative true answer. Essentially, this answerhood operator presupposes the existence of a maximally informative true answer p and picks out this p from Q, a set of propositions. What this answerhood operator does is reminiscent of the semantics of definite determiner *the*, which, when defined, contributes definiteness by picking out the unique (e.g., *the dog*) or the mereologically maximal (e.g., *the dogs*) item (see (10)).

(12) $$\textsc{Ans}(Q)(w) = \frac{\exists p[w \in p \in Q \wedge \forall q[w \in q \in Q \rightarrow p \subseteq q]].}{\iota p[w \in p \in Q \wedge \forall q[w \in q \in Q \rightarrow p \subseteq q]]}$$ Dayal (1996)

With the above two ideas combined, an interpretable *wh*-question can be analyzed in the same dynamic semantics framework as modified numerals and definite descriptions are analyzed in Sect. 2.

[3] I thank an anonymous reviewer for asking me to make this clear.

As illustrated in (13) (*wh*-movement and head movement are omitted in the tree), *who* works like *someone* or the indefinite component of *the* in introducing a dref in a non-deterministic way. After all relevant restrictions are added (here $\text{BOY}(x)$, $\text{KISS}(\text{MARY}, \text{x})$), the silent operator, $\mathbf{Ans}_u$, plays the same role as a maximality operator, bringing a post-suppositional evaluation and checking in the output information state whether u is assigned the mereologically maximal plural individual x that satisfies $\text{BOY}(x)$ and $\text{KISS}(\text{MARY}, x)$. Thus the application of $\mathbf{Ans}_u$ leads to a deterministic update. As far as a *wh*-question satisfies Dayal (1996)'s Maximal Informativity Presupposition and is thus interpretable, the derivation involving the application of **Ans** should not fail.[4][5]

(13) $\lambda g. \{\langle T, g^{u \to x}\rangle \mid x = \sigma x.[\text{KISS}(\text{MARY}, x) \wedge \text{BOY}(x)]\}$

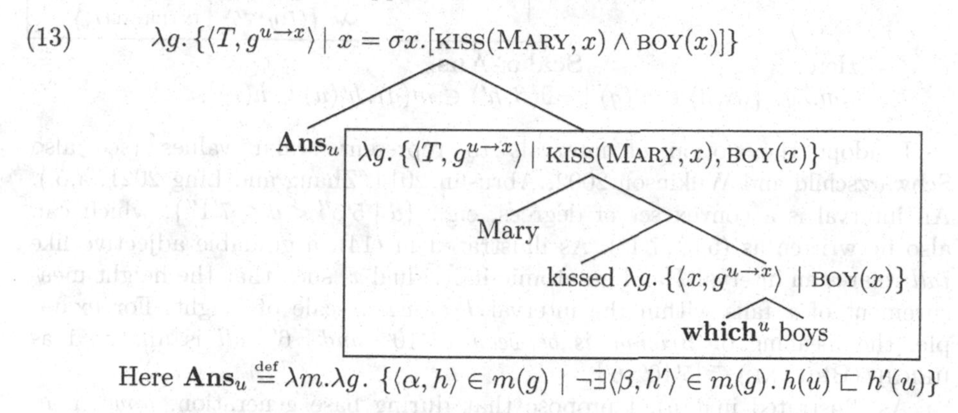

Here $\mathbf{Ans}_u \stackrel{\text{def}}{=} \lambda m.\lambda g.\ \{\langle \alpha, h\rangle \in m(g) \mid \neg\exists\langle \beta, h'\rangle \in m(g).\, h(u) \sqsubset h'(u)\}$

A similar analysis can be developed for degree questions, with an answerhood operator, $\mathbf{Scalar\text{-}Ans}_u$, which is adjusted for a set of drefs that are scalar values.

(14) $[\![\text{tall}]\!]_{\langle dt,et\rangle} \stackrel{\text{def}}{=} \lambda I_{\langle dt\rangle}\lambda x.\text{HEIGHT}(x) \subseteq I$ (Zhang and Ling 2021)

---

[4] In this short paper, I focus on the most basic data of *wh*-questions (e.g., *who did Mary kiss*) and degree questions (e.g., *how tall is Brienne*). I leave aside for future work cases like mention-some questions that can have multiple complete true answers (see (i)) or higher-order reading questions (see (ii) and Xiang 2021).

(i) Where can I buy an Italian newspaper?
(ii) Which books does John have to read?
The French novels or the Russian poems. The choice is up to him.

[5] Here I actually implement Dayal (1996)'s Maximal Informativity Presupposition in a post-suppositional way. As pointed to me by Anna Szabolcsi (p.c.), a post-supposition that is not satisfied in its local domain gets imposed on the non-local domain like a traditional presupposition (see Brasoveanu and Szabolcsi 2012), so maybe we could say that all presuppositions start out as post-suppositions. Thus I believe a post-suppositional version of Dayal (1996)'s presupposition is compatible with this general picture of presuppositions and worth having a try. A thorough investigation of this post-suppositional perspective on *wh*-questions and whether my current analysis is strictly contingent on this post-suppositional perspective is left for future work.

(15)

$\lambda g.\{\langle T, g^{u\to I}\rangle \mid I = \text{the contextually most informative } I \text{ s.t. } \textsc{height}(\textsc{Brienne}) \subseteq I\}$

- **Scalar-Ans**$_u$
- $\lambda g.\{\langle T, g^{u\to I}\rangle \mid \textsc{height}(\textsc{Brienne}) \subseteq I\}$
  - Brienne
  - is
  - **how**^{u} $\lambda g.\{\langle I, g^{u\to I}\rangle \mid \textsc{interval}(I)\}$
  - tall

Here **Scalar-Ans**$_u$ $\overset{\text{def}}{=}$ $\lambda m.\lambda g.\,\{\langle \alpha, h\rangle \in m(g) \mid \neg\exists\langle \beta, h'\rangle \in m(g).\, h'(u) \subset h(u)\}$

I adopt the notion of intervals to represent scalar values (see also Schwarzschild and Wilkinson 2002, Abrusán 2014, Zhang and Ling 2021, a.o.). An interval is a convex set of degrees, e.g., $\{d \mid 5'5'' < d \leq 7'1''\}$, which can also be written as $(5'5'', 7'1'']$. As illustrated in (14), a gradable adjective like *tall* relates an interval I and an atomic individual x, such that the height measurement of x falls within the interval I along a scale of height. For example, the meaning of *Brienne is between* $5'10''$ *and* $6'$ *tall* is analyzed as $\textsc{height}(\textsc{Brienne}) \subseteq [5'10'', 6']$.

As illustrated in (15), I propose that during base generation, *how*^{u} non-deterministically introduces an interval dref, I.[6] After relevant restrictions are added (here $\textsc{height}(\textsc{Brienne}) \subseteq I$), the application of **Scalar-Ans**$_u$ picks out the most informative interval from a set of possible intervals, leading to a deterministic update. Under an ideal context, where measurements don't involve any errors, this most informative interval would be a singleton set of degrees (i.e., the narrowest interval that entails all intervals satisfying relevant restrictions), containing the precise height measurement of Brienne (e.g., $[6'3'', 6'3'']$).

This post-suppositional view on the interpretability of *wh*-questions is also compatible with insights on (i) the cross-linguistic parallelism between *wh*-questions and *wh*-free relatives (Caponigro 2003, 2004, Chierchia and Caponigro 2013), and (ii) the categorial approach to *wh*-questions (see Hausser and Zaefferer 1979).

---

[6] (i) shows that *how* is parallel with other *wh*-expressions in introducing drefs and supporting cross-sentential anaphora. In (ia), *6′3″* is similar to definite descriptions or proper names (e.g., *Kevin* in (11b)) in introducing a definite scalar value so that *that* in the subsequent sentence refers back to it. Obviously, the parallelism between (ia) and (ib) is similar to that shown in (11).

(i) a. Brienne is $6'3''^{0}$ tall. It seems that Jaime is a bit shorter than that$_0$.
b. How^0 tall is Brienne? It seems that Jaime is a bit shorter than that$_0$.

As illustrated in (16), *wh*-free relatives can be replaced by truth-conditionally equivalent DPs, and in most cases (except for the complement position of existential predicates in some languages, see Caponigro 2004), both *wh*-free relatives and their corresponding DPs exhibit maximality/definiteness.[7] Under the current post-suppositional analysis, the semantics of the free relative in (16a), ⟦what Adam cooked⟧, can be derived by applying the silent maximality operator $\mathbf{Ans}_u$ to the meaning of the question *what did Adam cook?*, which yields the maximal sum of things, $\sigma x.[\text{COOK}(\text{ADAM}, x)]$, i.e., the meaning of the DP *the things Adam cooked* (see also Chierchia and Caponigro 2013 for a similar idea).[8]

(16) a. Jie tasted <u>what^u Adam cooked</u>. (example from Caponigro 2004)
b. Jie tasted [$_{\text{DP}}$ the^u things Adam cooked].

Within the categorial approach to *wh*-questions (Hausser and Zaefferer 1979), a *wh*-question denotes a function, which takes its short answer as argument to generate a (maximally informative) true proposition, as illustrated in (17).

(17) Categorial approach: ⟦who did Mary kiss⟧ = λx. Mary kissed x
a. Short answer: Kate and Kevin.
b. Propositional answer: Mary kissed Kate and Kevin.

Under the current post-suppositional analysis, as illustrated in (18), this function $\lambda x.$*Mary kissed* x is considered a restriction on the dref introduced by the *wh*-expression, x. Then the short answer, here *Kate and Kevin*, can be considered similar to the cardinality tests in the case of the cumulative-reading sentence *exactly 3 boys saw exactly 5 movies.* The test $\mathbf{(kate \oplus Kevin)}_u$ is attached to the application of the maximality test $\mathbf{Ans}_u$, checking whether $\sigma x.\text{KISS}(\text{MARY}, x)$ is equivalent to the sum 'Kate⊕Kevin'. This amounts to turning a short answer into a corresponding propositional answer to a *wh*-question.

[7] *Wh*-free choices corresponding to mention-some *wh*-questions are also exceptions (see Chierchia and Caponigro 2013) and don't seem to exhibit maximality:

(i) Mary looked for who can help her.
= Mary looked for <u>someone</u> that can help her.
≠ Mary looked for <u>all the people</u> that can help her.

[8] In addition to *wh*-free relatives, concealed questions also demonstrate the parallelism between definite DPs and *wh*-questions (see e.g., Nathan 2006):

(i) a. Jaime knows <u>how tall Brienne is</u>.
b. Jaime knows <u>the height of Brienne</u>.

(18)

$$\lambda g.\{\langle T, g^{u\to x}\rangle \mid x = \sigma x.[\text{KISS}(\text{MARY}, x)]\}, \text{ if } x = \text{Kate} \oplus \text{Kevin}$$

$$(\textbf{Kate}\oplus\textbf{Kevin})_u \quad \lambda g.\{\langle T, g^{u\to x}\rangle \mid x = \sigma x.[\text{KISS}(\text{MARY}, x)]\}$$

$$\textbf{Ans}_u \quad \text{who did Mary kiss?} \quad \lambda g.\{\langle T, g^{u\to x}\rangle \mid \text{KISS}(\text{MARY}, x)\}$$

Essentially, based on Dayal (1996)'s Maximal Informativity Presupposition, I propose that for an interpretable *wh*-question, (i) its *wh*-expression introduces a dref non-deterministically, and (ii) a delayed, post-supposition-like maximality operator can bring definiteness to this dref, leading to a deterministic update.

4 Accounting for Uninterpretable Questions

4.1 Interpreting a Modified Numeral in a Matrix Degree Question

The interpretation of a declarative degree sentence containing a modified numeral (see (2a), repeated here as (19)) is straightforward.

(19) Exactly three^u boys are between 5′10″ and 6′ tall.[9] (= (2a))

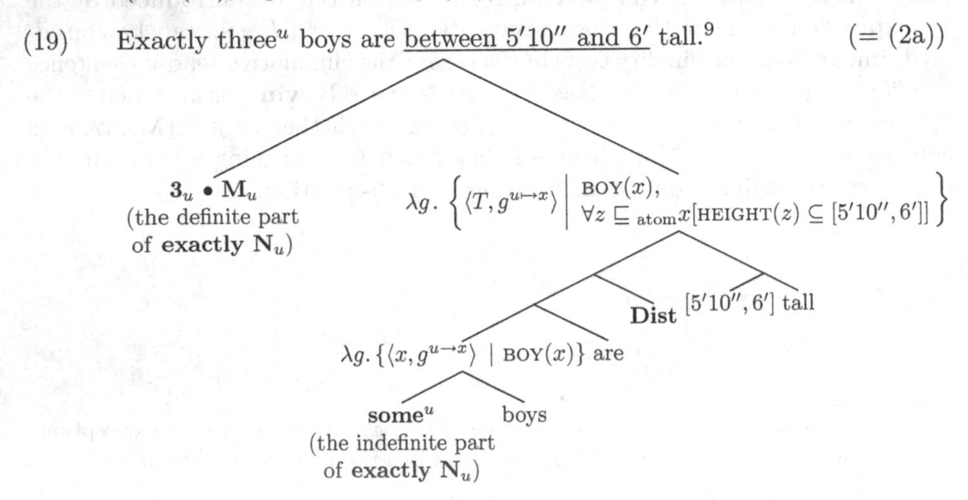

In this sentence, only *exactly three* brings post-suppositional tests. As shown in (19), as post-suppositional tests, $\mathbf{M}_u$ picks out the largest boy-sum x such that for each atomic boy within x, his height falls within the interval $[5'10'', 6']$, and $\mathbf{3}_u$ checks whether the cardinality of this boy-sum x is equal to 3.

Then I turn to the core data under discussion, a degree question containing a modified numeral (repeated in (20)):

---

[9] Given that $[\![\text{tall}]\!]$ relates an interval and an atomic individual (see (14)), I assume a distributivity operator **Dist** ($\overset{\text{def}}{=} \lambda x.\lambda P_{\langle et\rangle}.\forall z \sqsubseteq_{\text{atom}} x[P(z)]$) here.

(20) #How$^{\nu}$ tall are exactly three^{u} boys? (= (2b))

According to the post-supposition-based analysis addressed in Sects. 2 and 3, in sentence (20), both *wh*-expression *how* and modified numeral *exactly three* (*boys*) first introduce a dref, as show in (21):

(21) Before post-suppositional tests are applied:

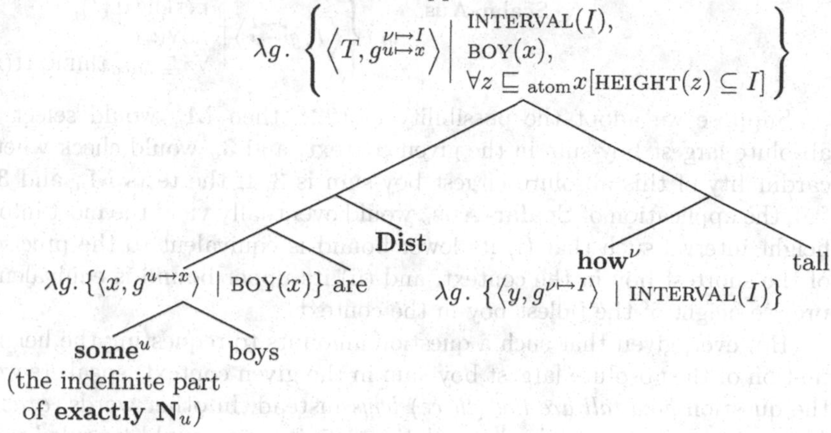

Once all the drefs are introduced and relevant restrictions are added, there are two potential derivation orders: either (i) as shown in (22), the maximality and cardinality tests of *exactly 3* are applied first, letting the deterministic update from **Scalar-Aus**$_{\nu}$ take place later, or (ii) as shown in (23), the deterministic update from **Scalar-Aus**$_{\nu}$ happens first, letting the maximality and cardinality tests of *exactly 3* be checked later.

(22)

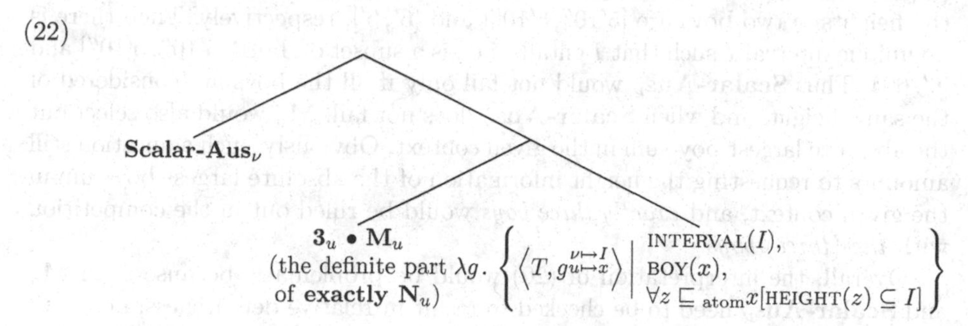

(23)

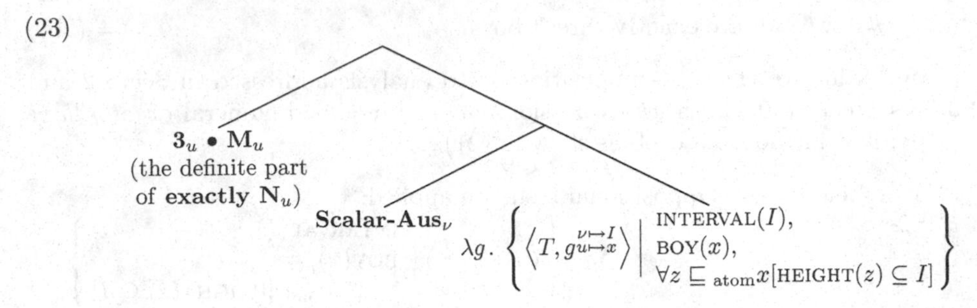

Suppose we adopt the possibility of (22), then $\mathbf{M}_u$ would select out the absolute largest boy-sum in the given context, and $\mathbf{3}_u$ would check whether the cardinality of this absolute largest boy-sum is 3. If the tests $\mathbf{M}_u$ and $\mathbf{3}_u$ don't fail, the application of **Scalar-Aus**$_\nu$ would eventually yield the most informative height interval such that (i) its lower bound is equivalent to the precise height of the shortest boy in the context, and (ii) its upper bound is equivalent to the precise height of the tallest boy in the context.

However, given that such a question amounts to requesting the height information of the absolute largest boy-sum in the given context, speakers would use the question *how tall are the (three) boys* instead. In other words, *exactly three boys*, which necessarily involves relative definiteness, would be ruled out in the competition with the definite description *the (three) boys*, which supports the meaning of absolute definiteness.[10]

On the other hand, suppose we adopt the possibility of (23), then **Scalar-Aus**$_\nu$ would select out the absolute most informative height interval such that it includes the height of some boy(s). As far as boys are considered not of the same height, there cannot be a unique most informative height interval (e.g., suppose the heights of two boys are $[5'10'', 5'10'']$ and $[6', 6']$, respectively. Then there is no unique interval I such that I entails, i.e., is a subset of, both $[5'10'', 5'10'']$ and $[6', 6']$). Thus **Scalar-Aus**$_\nu$ would not fail only if all the boys are considered of the same height, and when **Scalar-Aus**$_\nu$ does not fail, $\mathbf{M}_u$ would also select out the absolute largest boy-sum in the given context. Obviously, such a question still amounts to requesting the height information of the absolute largest boy-sum in the given context, and *exactly three boys* would be ruled out in the competition with *the (three) boys*.

Overall, the interpretation of (20) would be problematic because both $\mathbf{M}_u$ and **Scalar-Aus**$_\nu$ need to be checked to result in relative definiteness, i.e., both wait to be applied as the last post-suppositional test in the derivation. In other words, without resolving the update of u, **Scalar-Aus**$_\nu$ cannot work to yield the most informative height interval for u, but without resolving the update of ν, $\mathbf{M}_u$ cannot work to pick out the relatively maximal boy-sum. Obviously, the requirements of $\mathbf{M}_u$ and **Scalar-Aus**$_\nu$ cannot be both satisfied, and the unacceptability of the whole sentence thus arises.

[10] An anonymous reviewer asks whether this competition view would weaken the analysis, as it implies that the relative definiteness of *how* and modified numerals in *#How tall are exactly three boys* CAN be satisfied. In Sect. 4.2, I show that sometimes these post-suppositions that impose relative definiteness indeed can all be satisfied.

4.2 Interpreting a Modified Numeral in an Embedded Degree Question

As illustrated by (24), in comparative sentences, their *than*-clause can be considered parallel to a degree question (see Fleisher 2020, Zhang 2020).

(24) Brienne is taller <u>than Jaime is ~~tall~~</u>.
⟦than Jaime is⟧ $\rightsquigarrow$ addressing a degree question: *how tall is Jaime?*

According to Zhang and Ling (2021), a comparative sentence basically means that the scalar value associated with the subject minus the scalar value associated with the comparative standard results in a positive difference (i.e., an increase). As shown in (25), comparative morpheme *-er* is considered denoting a default positive difference, i.e., an increase. The *than*-clause, i.e., *than Jaime is* ~~tall~~ in (24), denotes the short answer to the degree question *how tall is Jaime* and amounts to the most informative interval I' satisfying the restriction $\text{HEIGHT}(\text{JAIME}) \subseteq I'$, written as $\iota I'[\text{HEIGHT}(\text{JAIME}) \subseteq I']$ here. Eventually, as shown in (25d), this short answer to the degree question *how tall Jaime is* plays the role of comparative standard in the derivation of sentential meaning.

(25) a. $[\![\text{tall}]\!]_{\langle dt,et\rangle} \stackrel{\text{def}}{=} \lambda I_{\langle dt\rangle}\lambda x.\text{HEIGHT}(x) \subseteq I$ (= (14))

b. $[\![\text{-er}]\!] \stackrel{\text{def}}{=} (0, +\infty)$ $\rightsquigarrow$ a default positive difference
i.e., the most general positive interval that represents an increase
(With a presupposition of additivity: there is a contextually salient scalar value serving as the base of the increase)

c. Assuming a silent operator that performs comparison:
Minus $\stackrel{\text{def}}{=} \lambda I_{\text{STANDARD}}\lambda I_{\text{DIFFERENCE}}.\iota I[I - I_{\text{STANDARD}} = I_{\text{DIFFERENCE}}]$

d. $[\![(24)]\!] \Leftrightarrow \text{HEIGHT}(\text{BRIENNE}) \subseteq \iota I[I - I_{\text{STANDARD}} = I_{\text{DIFFERENCE}}]$
$\Leftrightarrow \text{HEIGHT}(\text{BRIENNE}) \subseteq$
$\iota I[I - \iota I'[\text{HEIGHT}(\text{JAIME}) \subseteq I'] = (0, +\infty)]$

Intriguingly, although the matrix degree question #*how tall are exactly three boys* (see (2b)/(20)) is uninterpretable, comparative sentence (26), which contains a *than*-clause corresponding to the problematic degree question, is good.

(26) Mary is taller <u>than$^{\nu}$ exactly three^{u} boys are ~~tall~~</u>.

I have proposed an analysis for (26) in Zhang (2020). As mentioned in Sect. 4.1, for the matrix degree question #*how* $^{\nu}$ *tall are exactly three* ^{u} *boys*, **Scalar-Aus**$_{\nu}$ and **M**$_{u}$, both tests that bring relative definiteness, require to be applied as the last test, and both requirements cannot be satisfied at the same time.

For (26), however, information outside the *than*-clause contributes to resolve the deterministic update of ν independent of the update of u.

As mentioned above (see (25c)), the semantics of a comparative addresses the relation among three definite scalar values: (i) the scalar value associated with the subject, which serves as the **minuend**; (ii) the scalar value associated with the *than*-clause, which serves as the **subtrahend**; and (iii) the difference between the minuend and the subtrahend. Given the subtraction relation between these three definite values (see (25c)), we can use two of the three values to reason about the third one.

Thus, for (26), given the minuend (i.e., HEIGHT(MARY)) and the difference (i.e., $(0, +\infty)$), the deterministic update of ν (i.e., the value of the subtrahend) can be settled first: it is the largest interval below HEIGHT(MARY), which can be written as ($-\infty$, the precise height measurement of Mary).[11] This update of ν satisfies relative definiteness in the sense that it is checked at the sentence-level, beyond the *than*-clause itself, as a delayed test after more restrictions are added (i.e., μ is assigned to an interval dref I satisfying 'I is below HEIGHT(MARY)').

Then similar to the case of (19), $\mathbf{M}_u$ is applied to pick out the largest boy-sum x such that $\forall z \sqsubseteq_{\text{atom}} x[\text{HEIGHT}(z) \subseteq (-\infty,$ the precise height measurement of Mary)], and $\mathbf{3}_\nu$ is applied to check whether the cardinality of x is 3. Therefore, through the derivation of the meaning of (26), the relative maximality of *exactly three boys* is achieved.

It is worth noting that for this x, the interval ($-\infty$, the precise height of Mary) can still be the most informative short answer to the degree question *how tall is x* (i.e., with **Scalar-Aux**$_\nu$ applied to *how tall is x*). Imagine an extreme case: one of the boys in x is just slightly shorter than Mary is, and another one of the boys in x is extremely short. Then the application of **Scalar-Aux**$_\nu$ would lead to exactly this interval ($-\infty$, the precise height of Mary). In other words, the above analysis of (26) is not incompatible with the view that the *than*-clause addresses the short answer to a corresponding degree question. It's just that in this case, the information of this short answer (i.e., ($-\infty$, the precise height of Mary)) is derived first, and then this definite interval is made use of in checking the post-suppositional requirements of the modified numeral here (i.e., *exactly three boys*).

5 Discussion

In Sect. 4, I have shown that the uninterpretability of the pattern 'modified numeral + degree question' is essentially due to a conflict between different items that bring post-suppositions (i.e., both need to be applied as the last test to result in relative definiteness) and how this conflict can be circumvented (i.e., additional information is available to resolve the definiteness of one of the items and thus remove the conflict). Here I compare the current proposal with three existing lines of research on intervention effects or weak island effects.

5.1 Intervention Effects: Beck (2006) and Li and Law (2016)

Both Beck (2006) and Li and Law (2016) address intervention effects related to focus, but their empirical coverages are different. As shown in (27) and (28), their analyses target different problematic configurations.

[11] In our actual world, the height of a person cannot be a negative value. This should be considered a physical constraint in our world knowledge, not a linguistic constraint. Linguistically, we can imagine characters with a negative height in fantasy works.

(27) The problematic configuration analyzed by Beck (2006):
?* [*Q*...[focus-sensitive operator [$_{\text{YP}}$... WH...]]]

(28) The problematic configuration analyzed by Li and Law (2016):
?*
[...focus-sensitive operator [focus alternatives...ordinary alternatives...]]
(or ?* [...focus-sensitive operator [XP_F...WH...]])

Beck (2006) is based on Rooth (1985)'s focus semantics. A *wh*-expression has its focus semantic value (i.e., a set of alternatives), but lacks an ordinary semantic value. A *Q* operator is needed to turn the focus semantic value of a *wh*-expression into an ordinary semantic value. However, in the problematic configuration in (27), (i) a focus-sensitive operator blocks the association between the *Q* operator and the *wh*-expression, and moreover, (ii) the focus-sensitive operator needs to be applied on an item that has both a focus semantic value and an ordinary value, which the *wh*-expression lacks. Thus the derivation crashes.

According to Li and Law (2016), given that both XP_F and WH introduce alternatives, embedding WH within the scope of XP_F makes $[\![[XP_F...WH...]]\!]$ a set of sets of alternatives, which becomes an illicit input for the focus-sensitive operator, resulting in a derivation crash.

Both Beck (2006) and Li and Law (2016) explain the uninterpretability of intervention patterns as derivation crash. Different from these approaches, the current account for the uninterpretability of the pattern 'modified numeral + degree question' is based on a potential failure of achieving relative definiteness.

As shown in Sect. 4, for the pattern 'modified numeral + degree question', the potential failure of achieving relative definiteness exists for matrix degree questions, but not for embedded degree questions (i.e., *than*-clauses of comparatives). Thus empirically, the current account works better than existing approaches that explain uninterpretability as derivation crash.

It is worth investigating whether/how the current approach can be further extended to cover the data of intervention effects. As shown in (29), the matrix *wh*-question (29a) is problematic. Indeed, it has a problematic configuration in both the theories of Beck (2006) and Li and Law (2016). However, once this configuration is embedded in a *wh*-conditional, as shown in (29b), there is no longer uninterpretability. The acceptability contrast between (29a) and (29b) suggests that the problem of (29a) might not be due to a derivation crash.

(29) a. * zhǐyǒu Mary_F dú-le shénme shū?
only Mary read-PFV what book
Intended: 'What book(s) did only Mary_F read?' Chinese

b. Context: Only Mary is interested in the books I read and follows me to read them.
wǒ dú shénme shū, zhǐyǒu Mary_F (yě) gen-zhe wǒ dú
I read what book only Mary (also) follow I read
shénme shū
what books
'Only Mary follows me to read whatever books I read.' Chinese

Actually, the case of (29b) seems similar to embedded degree questions with a modified numeral (see (26) in Sect. 4.2). For (29b), suppose both the *wh*-item (i.e., *shénme*) and the focused part (i.e., *zhǐyǒu Mary* 'only Mary') introduce drefs first and bring post-suppositional tests later. Then within a *wh*-conditional, the deterministic update of the *wh*-expression can be resolved independent of the focused part, helping to circumvent the issue of which post-suppositional test needs to be applied the last. I leave the details of this analysis for future work.

5.2 Abrusán (2014)'s analysis of weak island effects

As mentioned in Sect. 1, the uninterpretable pattern addressed in this paper, i.e., 'modified numeral + degree question', is also reminiscent of weak island effects. However, the difference between (30) and (31) shows that the uninterpretable pattern under discussion (see (31a)) is still somewhat distinct from typical weak island effects (see (30a)).

In particular, weak island effects are observed for both matrix and embedded degree questions (see the parallelism between (30a) and (30b)), while the major pattern addressed in this paper only leads to uninterpretable matrix degree questions, but not to uninterpretable embedded degree questions (see the contrast between (31a) and (31b) and the discussion in Sect. 4.2).

The contrast between embedded degree questions in (30b) and (31b) is likely due to a further requirement of degree questions: the request of height information brings an existential presupposition for items undergoing height measurement – here only *exactly 3 boys*, but not *few boys*, guarantees that this existential presupposition be satisfied (see also Zhang 2020 for more discussion).

On the other hand, as illustrated in (30)-(31), it seems that DE quantifiers and modified numerals do not lead to uninterpretability in *wh*-questions that are not degree questions (see (30c) and (31c)).

Presumably, the contrast between (31a) and (31c) is parallel with the contrast between (30a) and (30c) and thus can be explained by an extension of accounts that aim to account for weak island effects. For the contrast between (31a) and (31c), a detailed analysis along this line is left for future work.[12]

(30) DE quantifiers in *wh*-quesions (Weak island effects: (30a) and (30b))

a. #How tall are few boys? — Matrix degree question
b. #Mary is taller than few boys are. — Embedded degree question
c. Which books did few boys read? — *which*-question (cf. (30a))

(31) Modified numeral in *wh*-questions

a. #How tall are exactly 3 boys? — Matrix degree question
b. Mary is taller than exactly 3 boys are. — Embedded degree question
c. Which books did exactly 3 boys read? — *which*-question (cf. (31a))

[12] I thank an anonymous reviewer for bringing up this issue, i.e., the contrast between (31a) and (31c).

Among existing studies on weak island effects, Abrusán (2014)'s account is also based on the idea that an interpretable *wh*-question needs to meet Dayal (1996)'s Maximal Informativity Presupposition. As illustrated in (4a) (repeated here in (32)), since there does not exist a maximally informative interval I such that $\neg\text{HEIGHT}(\text{JOHN}) \subseteq I$, (32) does not meet the presuppositional requirement, leading to uninterpretability.

(32) #How tall isn't John? (= (4a))

The current analysis is essentially in the same spirit as Abrusán (2014). Although Abrusán (2014) focuses on weak island effects, she raises the issue of how intervention effects and weak island effects can be connected. As addressed in Sect. 5.1, the current analysis has the potential of explaining intervention effects as well. It is also worth investigating whether the current analysis can eventually be extended to bridge between the phenomena of intervention effects and those of weak island effects.

6 Conclusion

In this paper, I have adopted a dynamic semantics perspective to explain why a degree question like *#how tall are exactly three boys?* is unacceptable. The account crucially relies on the ideas that (i) both *wh*-items (e.g., *how*) and modified numerals (e.g., *exactly three boys*) introduce drefs and bring post-suppositonal tests that result in relative definiteness, and (ii) when different post-suppositional tests are present, and their relative definiteness cannot be all achieved, uninterpretability arises.

Presumably, the current account will bring new insights on more empirical phenomena, in particular, intervention effects and weak island effects. How the current account will influence our understanding on the scope-taking issue within a *wh*-question is also left for future research.

References

Abrusán, M.: Weak island effects. OUP (2014). https://doi.org/10.1093/acprof:oso/9780199639380.001.0001

Beck, S.: Wh-constructions and transparent logical form. Ph.D. thesis, Universtät Tübingen (1996). https://hdl.handle.net/11858/00-001M-0000-0012-900B-9

Beck, S.: Intervention effects follow from focus interpretation. Nat. Lang. Semant. **14**, 1–56 (2006). https://doi.org/10.1007/s11050-005-4532-y

Brasoveanu, A.: Modified numerals as post-suppositions. J. Semant. **30**, 155–209 (2013). https://doi.org/10.1093/jos/ffs003

Brasoveanu, A., Szabolcsi, A.: Presuppositional too, postsuppositional too. In: Aloni, M., Franke, F., Roelofsen, F. (eds) The dynamic, inquisitive, and visionary life of ϕ, $?\phi$, and $\diamond\phi$: A festschrift for Jeroen Groenendijk, Martin Stokhof, and Frank Veltman, 55–64. https://festschriften.illc.uva.nl/Festschrift-JMF/

Bumford, D.: Split-scope definites: relative superlatives and Haddock descriptions. Linguist. Philos. **40**, 549–593 (2017). https://doi.org/10.1007/s10988-017-9210-2

Caponigro, I.: Free not to ask: on the semantics of free relatives and wh-words cross-linguistically. Ph.D. thesis, UCLA (2003). https://idiom.ucsd.edu/ivano/Papers/2003_dissertation_revised_2019-4-10.pdf

Caponigro, I.: The semantic contribution of wh-words and type shifts: evidence from free relatives crosslinguistically. In: Young, R. B. (eds) Proceeding of the SALT, vol. 14, pp. 38–55 (2004). https://doi.org/10.3765/salt.v14i0.2906

Charlow, S.: Post-suppositions and semantic theory. J. Semant. In Press (2017). https://ling.auf.net/lingbuzz/003243

Chierchia, G., Caponigro, I.: Questions on questions and free relatives. Handout of Sinn und Bedeutung 18 (2013). https://scholar.harvard.edu/chierchia/publications/demo-presentation-handout

Comorovski, I.: Interrogative phrases and the syntax-semantics interface. Kluwer (1996). https://doi.org/10.1007/978-94-015-8688-7

Dayal, V.: Locality in Wh Quantification. Kluwer (1996). https://doi.org/10.1007/978-94-011-4808-5

Fleisher, N.: Nominal quantifiers in than clauses and degree questions. Syntax Semant. **42**, 364–381 (2020). https://doi.org/10.1163/9789004431515

Fox, D., Hackl, M.: The universal density of measurement. Linguist. Philos. **29**(5), 537–586 (2007). https://doi.org/10.1007/s10988-006-9004-4

Haddock, N.: Incremental interpretation and combinatory categorial grammar. The 10th international joint conference on artificial intelligence 2, 661–663 (1987) Morgan Kaufmann Publishers Inc. https://www.ijcai.org/Proceedings/87-2/Papers/012.pdf

Hamblin, C.: Questions in Montague grammar. Foundations of Language **10**, 41–53 (1973). https://www.jstor.org/stable/25000703

Hausser, R., Zaefferer, D.: Questions and answers in a context-dependent Montague grammar. In: Formal semantics and pragmatics for natural languages, 339–358. Springer, Cham (1979). https://doi.org/10.1007/978-94-009-9775-2_12

Honcoop, M.: Towards a dynamic semantics account of weak islands. In: Galloway, T., Spence, J. (eds) Proc. SALT 6, 93–110 (1996). https://dx.doi.org/10.3765/salt.v6i0.2773

Karttunen, L.: Syntax and semantics of questions. Linguist. Philos. **1**, 3–44 (1977). https://www.jstor.org/stable/pdf/25000027.pdf

Krifka, M.: At least some determiners aren't determiners. In: Tuner, K. (eds) The semantics/pragmatics interface from different points of view, vol. 1, 257–291. Elsevier (1999). https://semantics.uchicago.edu/kennedy/classes/w14/implicature/readings/krifka99.pdf

Li, H., Law, J. H.-K.: Alternatives in different dimensions: a case study of focus intervention. Linguist. Philos. **39**, 201–245 (2016). https://dx.doi.org/10.1007/s10988-016-9189-0

Li, H.: A dynamic semantics for wh-questions. Ph.D. thesis, NYU (2020). https://www.researchgate.net/publication/348920343_A_dynamic_semantics_for_wh-questions

Nathan, L. E.: On the interpretation of concealed questions. Ph.D. thesis, MIT (2006). https://dspace.mit.edu/handle/1721.1/37423

Rooth, M.: Association with focus. Ph.D. thesis, UMass Amherst (1985). https://ecommons.cornell.edu/bitstream/handle/1813/28568/Rooth-1985-PhD.pdf

Rullmann, H.: Maximality in the semantics of wh-constructions. Ph.D. thesis, University of Massachusetts-Amherst (1995). https://scholarworks.umass.edu/dissertations/AAI9524743/

Schwarzschild, R., Wilkinson, K.: Quantifiers in comparatives: a semantics of degree based on intervals. Nat. Lang. Semant. **10**, 1–41 (2002). https://doi.org/10.1023/A:1015545424775

de Swart, H.: Indefinites between predication and reference. Proc. SALT **9**, 273–297 (2006) https://dx.doi.org/10.3765/salt.v9i0.2823

Szabolcsi, A., Zwarts, F.: Weak islands and an algebraic semantics for scope taking. Nat. Lang. Semant. **1**(3), 235–284 (1993). https://www.jstor.org/stable/23748421

Szabolcsi, A.: Strategies for scope taking. In: Szabolcsi, A. (eds) Ways of scope taking, 109–154. Springer (1997). https://doi.org/10.1007/978-94-011-5814-5_4

Szabolcsi, A.: Strong vs. weak island. The Blackwell Companion to Syntax, vol. 4, 479–531. Blackwell (2006). https://doi.org/10.1002/9780470996591.ch64

Umbach, C.: Why do modified numerals resist a referential interpretation. Proc. SALT **15**, 258–275 (2006). https://dx.doi.org/10.3765/salt.v15i0.2931

Xiang, Y.: Higher-order readings of wh-questions. Nat. Lang. Semant. **29**, 1–45 (2021) https://doi.org/10.1007/s11050-020-09166-8

Zhang, L.: Modified numerals revisited: the cases of fewer than 4 and between 4 and 8. Proc. Sinn und Bedeutung **21**(2), 1371–1388 (2018) https://doi.org/10.18148/sub/2018.v21i2.204

Zhang, L.: Scopelessness and uninterpretable degree questions. In: Bekki, D. (eds.) Proceeding of the International Workshop of Logic and Engineering of Natural Language Semantics (LENLS), vol. 16 (2019). https://www.academia.edu/43799952

Zhang, L.: Split semantics for non-monotonic quantifiers in than-clauses. Syntax Semant. **42**, 332–363 (2020). https://doi.org/10.1163/9789004431515

Zhang, L., Ling, J.: The semantics of comparatives: a difference-based approach. J. Semant. **38**, 249–303 (2021) https://doi.org/10.1093/jos/ffab003

Zhang, L.: Cumulative reading, QUD, and maximal informativeness. In: Proceeding of the LENLS, vol. 19 (2022). https://ling.auf.net/lingbuzz/006858

Author Index

D. Deng et al. (Eds.): TLLM 2022, LNCS 13524, p. 189, 2023.
https://doi.org/10.1007/978-3-031-25894-7